SECULAR TIME
and
THE NORWEGIAN SOMMAROY QUESTION

SOME OTHER WORKS BY SAMUEL K.K. BLANKSON

The Metaphysical Foundation for Physics
Why Time is not a natural Phenomenon
The Einstein Theory of Space-Time Without Mathematics
Time in Science and Life---the greatest legacy of Albert Einstein
How religious scientists play down the greatest of Einstein's achievements
The coming revolution in Physics
Time and the Application of Time
The logic of Time in the Universe
Philosophical Essays
On the Nature and Passage of Time and 4-D Geometry
Past, Present and Future as Time in the Age of Science
What is Time...?

First published in Great Britain in 2019 by
PRACTICAL BOOKS

Published by Blankson Enterprises Limited

London

www.practicalbooks.org

ISBN 13: 978-0-244-23645-8

A CIP catalogue record for this book is available from the British Library.

CONTENTS

FOREWORD/PREFACE

*In this book the Norwegian question as conceived by the author refers to the query, "What is Time since there is only one day in astronomy?" But the best expression of the idea is that there is only one day **in time.** There is only one day because the nights are just shadows of the planets going round the sun and are, as shadows, cosmically irrelevant.*

The recognition of this state of affairs automatically makes time secular in origin as it is calculated from only one day, exactly as physically perceived or practically discovered by the clever inhabitants of Sommaroy in Norway, all of them down to one man, and even all the people of Norway, for being so lucky to live in that privileged position and clever enough to know that it has serious implications for the study of time, for after all time controls everything in the universe of human knowledge. Philosophically Norway appears to be the luckiest nation in the world----certainly the most contented as I see it.

The literary, philosophic or scientific aspect of time is another matter. To be honest, the book was finished as "The Origin of Secular Time" and ready for the printers when the Norwegian question came up in June 2019.[1] So it had to be revised to take account of Norway's original,

---

[1] This is the practical proof of the Einstein notion that there are as many time systems as there are planets due to the discovery of local time. I took it to mean that Einstein's theory of time as a secular entity has been physically confirmed by the people of Norway, and changed the book's title. Later I pleaded with the Norwegians to join me in telling the world what we have found, thanks, of course, to Albert Einstein's adoption of the Lorentz local time concept as argued in the book---namely, the Lorentz local time idea is important but not Time Dilation which seems to distort physics by claiming that time behaves magically. Secular time is discrete because it is produced in units by the use of

unprecedented, great, lovable and clearly revolutionary idea of time--in fact, everything we want to know about secular time. This is the content of the entire book in a nutshell, namely that there is only one day in the universe and that the nights and darkness we experience are mere shadows of no consequence in astronomy or science---in the same way that a person's shadow does not affect our planet in anyway. That life would not exist without the nights is a matter for the biophysicists to discuss and must not be used to confuse the debate as to whether or not time is secular, because, after all, the life is not recognised by the cosmos. The two aspects of reality come into play here: there is human reality on earth and material reality in interstellar space where we're not even known to exist.

Time, Speech and Language are the real mysteries of the universe and all existence, as they are the essential ingredients of all knowledge. Logic, science and mathematics---taken as 'rational thought'---come next in order of importance, yet the universe itself has no idea of 'constructed' time (or time sequences) as we have here on earth, as it has no idea of our speech and languages. Otherwise, constructed by whom? As opposed to the discredited general, fixed and absolute divine time, secular time, limited to a planet, is conceived by man whose creations are not regarded in the universe as having any effects on anything. This vast universe, obviously, subsists on chance, chemistry, and mere being-

---

points. Thus discrete time cannot move faster or slower according to the conditions of space. It is not affected by space. As it is produced in units, time is relation between points, as Bertrand Russell stated in his Analysis of Matter. Let me stress (as I am going to do all through the book) that time does not exist independently to be affected by space---it is rather produced by space as relation between points when applied to space, any space, so that anybody anywhere can produce his 'local time', simple. This is the idea by which Einstein abolished divine time, to distort it with time dilation and the rest of it is sheer humbug.

--just being there.[2] Yet scientists will never agree that objects do not have time in them inherently simply because they occupy any space, implying that time is naturally present in every space and therefore is spread through the cosmos.[3] This is what is meant by the term 'space-time', and secretly they probably think Minkowski, the inventor of the theory (known as 4-D geometry) is next to God, particularly because Einstein praised him, which I think was just a crafty move by a very clever man.[4] Luckily science is open and so if and when our leading thinkers and researchers come round to realise the error of their ways

---

[2] It is about time we realised that human intelligence is unique. We don't know how we got it, but we have, and it is unique---other Beings will have their own unique attributes and most likely will be different from ours. Why this is so we don't know; all we know is that this intelligence will not last, and is never trouble-free either; as such intelligence is probably an accident as well. That it's never trouble-free is one of the curses of life we simply have to endure. But the reasons things don't last is the chain of causality and the energy that powers it--- everything is caused by something's duration that is also caused by some other thing's duration and so on, **but to man duration means time.** All of these are accidental; therefore life must have come about through an accident. Being an accident, it may and yet may not be replicated elsewhere. It is my hope that more secular time researches will uncover greater details about the origin of life, simply because life is controlled by time, **and time is duration, caused by some physical event or events. Yet duration means during the existence of something. So if duration is time then time is existence as created with mathematics to suit mechanisation in the clock. Ergo, mathematicians created time. Understanding this stretches the tether of you mind without increasing your knowledge of the nature of time---until we know the secret of life, for time is existence. All the mathematicians have done is to enable us carry portions of existence with us in the clock or watch.**

[3] If only they'd realise that this takes them back to the concept of deity they decry.

4 We can retain the term space-time (by the 3+1 formula in mathematics), but not in the sense that space and time constitute one entity, as argued all through the book. I have made time my special subject, and predict that physics is going to face serious problems with their current attitude about time and space-time.

about this matter, the problem of time will receive the attention it deserves, since it is the most intricate in the world. The purpose of this book is to demonstrate that it is only through Secular Time that the problem of the passage of time can be resolved, even though some writers insist that time does not exist and still proclaim that time travel is 'a scientific possibility'. That is what the strange thing called 'time' can do to some writers' train of thought. I hope this convoluted way of introducing my own arguments even before a long formal Introduction will be granted The Royal Pardon through the indulgence of our friendly critics.[5] I know they can be very charitable, especially those attached to the religions---such saintly critics, unfailingly rational, truthful and fair, but often cast as worst than the devil. In any case, it should be noted that all the religions rely on time without knowing what it is. I must beg the reader to bear with me, because the main wisdom of the book is in the Preface, the Introduction and Conclusion; the technical aspects are in the text, Footnotes and Appendixes. It may not be the best planning for a book, but I will always plead for leniency because this is about time in its basic logic. Nobody knows what it is and therefore how to write it out. I even believe that no one, single, individual should write such a book, unless, like Bertrand Russell, he's a genius---yet none of the scholars I approached to join me even bothered to reply. They probably thought I was mad. A literary agent kind enough to reply asked me what it is all about and why it is important. Yet if man is going to understand life and time the knowledge will be new and convoluted like this. It will

---

5 A number of major books have used this format and so it is nothing new, but all through the book the organisation is peculiar due to the nature of the subject, for time is very strange. In fact the problem of time is our greatest intellectual puzzle. But the original mathematicians who created it were so good that time, as we know of it, can either be explained with science or religion making life difficult for the professional philosophers.

not come from traditional sources---because they are traditional, have been there for ages, and hadn't solved the problem. By the way, time is important because time controls activity, and without activity we couldn't create civilization and maintain it.

To begin with, the main aspects and implications of time the book is about (among others) are these: The passage of time; the notion that time flows or is passing through nature (which is implied in so many statements of time); the idea that history is the march of time; and the theory that time just happens to be there, that it 'just is'---see the long footnotes to Part One for part of the gist of my theory (or my ABC) of time and existence, since time is the kingpin of all 'Being'. But I argue that 'Being' on its own is not time. It is chemistry. Being cannot be mechanised in the clock to be ticking time away. Motion is not time either, because time requires points---as the intervals between points or events. Clearly these can be mechanised in the clock.

Fortunately, although it is infinitely complex (and often wrapped in divine mysteries), time can be analysed in logic and logic is the ultimate arbiter of truth, not science, philosophy or physics, for these have to be grounded in logic and the data of sight or fail.[6] However, due to the atomic and nuclear means of destroying the world which happen to be the foremost subjects in physics (as the merchant of death), physicists claim that certain matters connected with time must be explained to their satisfaction or they will abide by the age-old religious notion of time as 'just is', meaning it just happens to be there, even though they insist on everything being necessarily conceivable under QED. They are Time and Gravity, Time and Entropy, Time Dilation, The Twins' Paradox,

---

[6] Thankfully thinking is free. We need no licence to think and do logic even to the embarrassment of the authorities in civilised countries.

The Origin of Secular Time

The passage of Time, and the strange behaviours of the quantum as regards time.[7]

The last query must be answered immediately before the quantum, being the magical particle it is (the smallest piece of matter in human experience), makes me forget it. It is that the quantum does not belong to this world; it may have helped to create it but since we humans have then gone on to construct our own different world, the quantum is alien here and cannot be controlled by this new world of mankind---the culture, roads, cities, ships, planes, etc. These are the worlds of the atom not where the tiny and almost immaterial quantum can exercise some influence. Sometimes a dose of ordinary commonsense can be helpful even in philosophy, and however time is defined it ends up containing a huge dose of human interference to say the least, not something the immaterial quantum would know how to handle. That, in my opinion, is the reason we're having problems with the quantum and time, especially secular time, and the fact it can appear in more than one position at the same time---of course, because it is not its time; it's unknown to it. This should make us ponder 'a world without time and order'. The human imagination will certainly struggle with that, in the same way as the quantum is struggling with our regime of order and

---

[7] The innocent reader may query why we should obey the physicists. The short answer is that physics is the chief merchant of death in all the complexity and vastness of the universe. It knows best the ways of putting materials together for destruction. Leaving physics to struggle with the interpretation of time (which controls everything), is to pave the way to metaphysical doubts, ignorance and religious internecine. In fact this is also the very reason why rational philosophy as championed by Bertrand Russell, A.J. Ayer, Sir Karl Popper and all those against The Bomb is important. Yours truly is simply following in their noble footsteps, but my experience is that probably human beings are really not worth saving.

strict time sequences based on the orbits of the sun, and so strict because nobody can reach up there to interfere with them. The Silicon Valley mavericks who are bent on destroying the world with dangerous microchips haven't come to that yet. But the strongly held believe that time just exists in the cosmos and is the same everywhere as well as being present in the nature of everything is the silliest in all science, since the Minkowski equation upon which it is based is clearly flawed. You will not find these views in the textbooks of physics only in the books dealing with the philosophy of science, or of physics, as scientists have to realise that the theory of knowledge in philosophy, epistemology, covers all scientific subjects as well.

Another contentious issue we must get out of the way is the 'scientific' time travel said to originate from Albert Einstein's ideas. Under the heading, "Unravelling the puzzle of travelling back in time", The Sunday Times (18th March, 2018, p.24), printed this: "After Einstein---who showed that space and time were curved---this began to seem possible..." Yet the whole statement is a bogey nonsense. Ordinary space is not in any way curved, and Einstein never said anything even remotely like that. What he proved is the cause of gravity, which he showed to be caused by the curvature of space as a result of the presence of a massive object. Hence Professor Eddington's delegation went out to prove the bending of light in a gravitational field.

But, as explained below, it was Hermann Minkowski who claimed to have equated space to time with mathematics, and the usual suspects began speculating that when space curves it will take time with it, and therefore time travel is 'a scientific possibility' for you to meet your grandparents before they were married----such nonsense. There should come a time (soon) when peddling such falsehoods becomes a criminal offence for causing unnecessary panic. I wonder what the IOP is going to do about that. Though the term 'space-time' can be used in the 3+1 sense, not in the sense that space has been (physically) equated to time-‑‑‑physics would become very easy if that were so and reality would not be as we have known it. If man has to fear anything it is not God but physics, the physical reality that runs through us, the world and the universe and whose real nature we are approaching but only slowly and very cautiously.

For the rest, let us begin with the logical definition of time so that we can agree on what we are talking about, for one of the sources of conflict in the religions is that no two sects can agree on what nature or man's destiny is or how the universe was created, with one of their leaders even claiming that God created the universe exactly at noon on AD 4004. Time in logic is 'relation between points', according to Bertrand Russell, and I agree with him.[8] It covers duration to any length, either serially or in one go. One reason is that this has always been the un-stated definition even among the religions‑‑‑ie, the year is one unit of time between points; so also is the second. Of course they call it 'just is', but when that is analysed, it becomes intervals between points (a sense of waiting between points), the same thing as 'relation between points'.

---

[8] The physical nature of time can never be known because it does not exist; time is conceptual and we can think of it in the following manner. The tram, for instance, cannot be running on the same track several times in a day or even an hour, because each trip is separated by a time lapse, known in mathematics as 'a time coordinate or the time coordinate'. In linguistics this means a gap or an interval, during which anything could have happened to the tracks down to the atomic level. So time is a gap, and a gap is a time coordinate. The point is in all activities there will be unavoidable gaps between events or points; so time is intimately associated with activity. Without activity there is no need for time, nor even the sense of time and timing.

If we reject Idealism, as we must when science and logic are mentioned (especially German Idealism from Kant all the way down to Wittgenstein),[9] then the cultural use of time, the thousand-and-one ways of 'naturally' using intervals between events, both in thought and action, are the instinctive use of the sense of waiting between points or events and must not be confused with the logical, scientific and philosophical definition of time which goes to the root of human existence and intellectual pursuits. It has been noted, however, that that tendency and social practice make the logical study of time very difficult, as those who are not professional thinkers cannot see the point of it. When religious thinking interferes, then questions about time are regarded as moral crimes. Yet still I think questions about time will force changes in physical theories in the future simply because it just is not just there, nor part of space---chemistry, brain

---

[9] This subject does not deserve a chapter because Idealism is the philosopher's magic wand by which every problem is soluble. I do not think or write like that. It is religion by another name. In order to advance the philosophy of science or the rational course of humankind, there are three books waiting to be written. Sadly I am too old to attempt them. They are Science v Idealism---which is true? Reality v Idealism and Life v Idealism, for it's always rational thought as against religious ideas disguised as technical philosophy (saying just what we want to hear), gleefully published by illiterate publishers and hailed by Oxbridge and the like as the greatest philosophy of all time. Sir Karl Popper was right to feel ashamed as a philosopher, not so much as a scholar but because charlatans have come to dominate the field---German Idealism and Wittgenstein, for instance. (German Idealism inspired Nietzsche and Nietzsche produced Hitler), Existentialism is another disgraceful stigma. I will never cease wondering why nobody is brave enough to ask Oxbridge why Bertrand Russell never even once mentioned the name 'Wittgenstein' in The History of Western Philosophy. Wittgenstein wanted to destroy physics. Can any sane thinker aspire to do that? As a non-European, or non-White, I am giving the Europeans this free advice about how we see them---hate and wars and attempts to exterminate your neighbours are not the purpose of the numerous Churches.

function, cultural practices, inertia and random actions from many events are all involved. And how did puny man invented this complex system of thinking about the world, life, and even the universe?

So let me state for all time (without fear of contradiction in rational thought) that time in logic, science and philosophy is a mathematical device by which man apportions manageable units of reality, duration (which means 'during the life of something') and existence or daylight for long and short specific tasks. But a complex mathematical device of this magnitude could only have been created long after abstract mathematical thought was invented. It has to be noted, though, that this definition may be just our human time for action, planning and control. Other forms of time may be caused by delayed action, chemistry or accidents in nature, particularly in the cosmos at large outside human control. Things will always happen in the absence of mankind, and they would betray signs of temporal intervals. The reason the public generally can ignore the philosophers about time, so much that Einstein's theory of time is unknown. This makes the analysis of time difficult as some events may be seen in contrary ways scarcely amenable to our standard theory of interpretation. However, in logical thought related to the earth, the definition given above is accurate. At least 'The Norwegian Question'[10] mentioned below confirms this interpretation of secular

---

[10] See "Time's up for Islanders who want to stop all the clocks"---The London Times, 19th June 2019, pp32-33. I call this The Norwegian question for logic, science and philosophy. In fact, I mean to say it is the basic question for all life on this planet and requires answers from logic, science and philosophy: in logical thought there is not much to do. Einstein has already solved the problem by declaring, correctly, that "There are as many times as there are inertial bodies", averred by Bertrand Russell that "There is no longer a universal time..." The scientific problem, on the other hand, is rather simple as it becomes a mater to be decided by material reality or the data of sight which proves that there is only

time---namely as our own device for the beneficial control of nature, affairs, people, events or every facet of reality.

Let me stress, however, that the notion that time is just is, or just happens to be there, survived relativity. Einstein was never able to decide on the provenance of time. Rather it is Bertrand Russell (that is why I tend to put them together) who said that, because of relativity, time is 'a construction' in his book Mysticism & Logic. I agree that it is a construction by man, because it is obvious to me that to create the SI of time (as a unit of reality, duration or existence), the space traversed by the earth round the sun had to be sub-divided cyclically (every 20 kilometres or so is equal to one second in earth time so that 31,536 000 seconds amount to one full circle of the sun, and we start another year.) I contend that this is the ultimate scientific, philosophical and mathematical provenance of time.

In plain language, time is not different from existence; it is the same thing as existence, but it is in units. So, essentially, time is existence broken down to culturally manageable units we can carry with us---as in the clock or watch. To achieve this we have had to apply points to sub-divide reality, the data of sight, or existence. So every unit of time is a piece of reality, otherwise we could not order people to perform tasks

---

one day in astronomy, and that the nights are mere shadows of no consequence----there is nothing we can only do by night and not by daylight, which philosophically means time is secular and that all the proclamations of all the religions about time are fallacious. We know, of course, that we owe our lives to the night periods, but that has nothing to do with philosophy as nobody planned it. **And yet that is the all important point, as it shows how things can come to be without planning---examples are the human, animal and plant lives. So, analysed in philosophy, the Norwegian question becomes the human question.**

within certain periods---otherwise, periods of what? The answer, of course, is periods of reality or existence; furthermore these periods can differ, ten minutes is longer than five minutes in duration, all of which have to be related to, or based on, reality to make sense.

Everything depends on the use of points. Without points you cannot have intervals; and without intervals there is no time, for time is from when to when.[11] But since points were invented by man, this makes time secular; yet people have gone to wars to defend time as divine. All the religions are exactly like toddlers, often fighting over things about which they know nothing.

What the mathematicians did with time as just is, is the same thing we are now doing with time as relation between points. It is based on the SI of time; and that makes it discrete. It means all time is known and used only in units, consisting either of multiples or fractions of the second as our SI of time. But time in units can only pass by in units; and since we use time to live and grow, it means the time thus created in units is (or can only be) absorbed in units for living and growth. It goes nowhere. This logical explanation of time cannot be refuted, no matter what mathematical big guns are deployed. Mathematicians are in the habit of assuming that mathematics is divine, with them as the noble messengers of the Gods; but Bertrand Russell showed that they're mistaken. The subject is speculative, tautological and frequently wrong or our maths geniuses don't even know what they are talking about, and

---

[11] Yet these things have to be planned or constructed, that is the reason there is no time in the cosmos and all action occur through accidents, random action or force, motion, etc. Thus we can use logic to back the assertion that there is no time in the cosmos, or we can go round the cosmos to find out, after all we'll come back younger than before!

that any intelligent person forced to study the subject at an early age can become proficient at it---above all, most of the formulas they build upon may be out of date, even plainly wrong. For instance, the 'S=ct...' equation is carried very far into scientific literature, yet it is completely false. In my opinion Minkowski was a distortion not a boon to relativity. Yet he is so honoured---more for the complexity (not the truth) of his mathematics.

Let me stress one essential point: an example of time is the daylight. We have divided it into twelve hours. How did we get the hours and how do they advance from one to twelve? These are the basic questions we want any theory of time to explain. The answer is that we count cycles. The mathematicians have done a great job here and must be commended.

This is what we see when logic is applied to the study of time. Even the Nobel Prize winner, Professor Richard Feynman recognised this in his physics lectures, for when physics is advanced to the level of theoretical physics and astronomy, it begins to resemble philosophy.[12] Physicists may think they're right to despise philosophers, yet that is the truth: legitimacy in science is obtained from logic and philosophy. The control of human knowledge began as theological tyranny from which scholars struggled to escape. But emancipation has now given scientists the notion that they can investigate and do anything as the inevitable course of human knowledge even if it meant the total annihilation of the human race. This is wrong. There is good knowledge and bad knowledge. Man needs to pursue only what is good for the human race as a whole and philosophers using logic is our best hope.

---

[12] Feynman, The Lectures, 17-1 to 17-3.

For instance, now that we know diesel is dangerous for the lungs, who can argue that it is cheap and therefore diesel vehicles and fuels are what should be produced?

Anyway, Using the earth's long orbit of the sun may not seem like counting cycles, but it has been cleverly organised (divided mathematically) to make it do exactly that---about twenty kilometres is equal to one second on earth; so all units of time are either multiples of seconds or fraction of it, <u>thus I infer that there is only one second, only one year and only one day</u>.

It means time is constructed by mathematicians. By counting seconds every 20 or so kilometres, we get the round figure that equals exactly to one orbit of the sun for us to start another year. That is how we get our time, essentially as "the manufactured instinct in the human mind required for living safely on the planet." This took centuries and a lot of mathematical thinking went into it (and still going on), since the sun is the source of all the energy we use and need. It seems mathematics can mould our brains and human nature in some way. Certainly mathematics is the rarest and most essential talent on earth, for basically it is thinking at its best and most profound, accurate, metaphysical, et al.

How does this time pass through nature? The answer is that if we can ignore the legends then time does not pass by physically, because it is only a concept we use to manage reality for human convenience. Otherwise time has nothing to do with reality, and it is wrong to assert that things change 'over time'.[13] The time is only coincidental---

---

[13] Daylight, for instance, is reality and we estimate that it lasts for twelve hours; but that has nothing to do with the daylight, or the reality. How we do our

because there is always time. In fact, changes occur through chemistry and other factors. It is completely wrong to say time causes all changes, including entropy. Religion bears much of the blame. People believe the religions; yet they know nothing of science, astronomy and human nature beyond the urge in man to worship. If man is not so fragile at birth and during infancy, this dependency on 'a father figure' would not make religion so popular.

Anyway time has no medium of transmission to go anywhere. It is only conceptual and not in anyway physical to have its own medium of transmission. It also has no power to cause any action; everything that happens can be explained in physics or by the laws of the physical world, especially under QED. Secular or discrete time is the same as reality, but sub-divided with mathematics into miniaturised units of reality we can carry about with us for cultural convenience---e.g. a one minute picture is the same as one minute reality of the world, except that it lasts for only a minute. So the mystery of life does not reside in time; time merely reflects it. Also we've never had 'A Brief History of Time'. It's impossible because time is expended in usage and therefore has no history---and as I will repeat again and again, history is not the march of time but the march of events; the time is associated with the events as they (the events) march on to the future.[14] The future is the future of

---

estimation (counting cycles, measuring relations between points and calling it time or whatever), is our own business; it has nothing to do with the reality out there. However, it seems we are able to do our estimations so successfully that we do not go out into the bush at midnight to get hurt.

[14] So many writers think nothing about conflating time with existence so much that another writer also wrote, "The moon's timescape has no flow", instead of (probably) just trying to say the moon has no air or wind flow. All this, more than one hundred years after Sir Arthur Eddington has said---because of Einstein's ideas about time---all those talking of 'time flow' are making

events not time; time is immaterial. Mathematicians tend to make of time whatever they want, but they are wrong. In philosophy time cannot march. It is immaterial and would not exist at all but for the fact that gaps appear in all activities----time is constructed from the gaps (intervals) in activities or between points. In the absence of activity, time disappears. Even the need for time will not arise.

It is assumed that originally there was no time at the inception of life. Yet man cannot live without activity; so life and time (meaning the need for activity and therefore time as 'the activity regulator') are intertwined. However, since time requires points and arithmetic, it was obviously invented by the brain as a complement to life. Otherwise the universe itself has no time or sense of time; for time requires sense and yet the universe has no sense of its own in 'central control' of action. This is the concept of secular time upon which this book is based. It is not enough to linguistically refute the religious view of time; it is necessary to show the logic of the secular challenge. Religion is not useless; in fact many people need it to live as a legal requirement. The problem is that life is full of problems, most of which can only be

---

meaningless noises...So I conclude that relativity is still not properly understood! I suggest that the kinematic sections of the June paper should be revisited. For relativity is a comprehensive epistemological package; to despise philosophy and philosophers as intensely as almost all scientists tend to do, is not the best attitude for studying the ideas of Albert Einstein, history's greatest philosopher/scientist. Countless papers about worm holes, politically different worm holes, a thousand dimensions of reality, the demise of time in black holes (as was once done to the eather),are never going to defeat Einstein, the one who killed the eather debate. They will only succeed in making physicists sound dazed and bemused, As stated in the text, they're now saying Einstein didn't even understand 4-D Geometry, and I believe them. Furthermore, curved space cannot take time with it to make time travel 'a scientific possibility' because the Minkowski equation of space with time is flawed and logically untenable.

resolved by thinking logically, and this logical thinking tends to clash with religion. When that happens about any matter, then we demand the logical grounds of the challenge. Because religion came first, to ask man to dispense with any aspect of it requires convincing arguments. We know this causes fracas because religious people tend to be unreasonable to the point of needless violence. We never hear of scholars throwing stones at each other over any issue; but different religious sects are doing it all the time round the globe.

We eventually learned to invent time to regulate activities; the reason time and life are so closely bound-up together in the mind, for it is the creation of the human brain, the part of the brain that invented mathematics. Naturally human beings marvel at the powers of the human brain; yet, as related to the size and complexity of the cosmos, the brain is like an ant's pee compared to the mighty oceans all put together---won't even register an effect! Man, I am afraid, does not count in the universe which supports him, which is an irony. The creation of human life appears to us as a mistake that would constitute the offence of manslaughter in law.[15] All women should be advised about this before they get pregnant. A time will come when giving birth will require a licence. There is nothing in life that is permanently good. There is no logic in life. If life has to end no matter what, then how can it

---

[15] The one insoluble conundrum of life is that man is so small yet he sees the cosmos as so big. Is it due to our size? It could be because the human toe alone would seem like a mountain to an ant. Let us infer from that and compare the knowledge with our assessment of the size of the cosmos. I am throwing this query to the mathematicians. Man is so small in comparison to the cosmos, at the same time, even as small as he is, he can affect minor changes in the universe--- eg. we can make enough atomic bombs to pulverise the earth out of existence thereby altering the gravitational effects among the planets---why is that? To my mind, religion is not the answer, since it can't even control the sinful popes.

hold anything good for mankind, Kings and Queens? Sometimes at moments of success in life, death strikes unexpectedly. Also, at every stage of human development, we hail new inventions only for them to turn problematic or dangerous---the motor car, aircraft, fossil fuels, the chainsaw, general science, chemistry and bombs, etc. etc. Even the useful telephone can be used for good or evil, and what about the computer and internet? So long as there is no pre-ordained destiny accepted by all, man has no sense of permanent goodness and one man's meat will continue to be another's poison---which is the guarantee that disagreements, conflicts and wars will never end. Once we're all going to Heaven; then we're told that there is no Heaven anywhere, and pointing to the sky is useless since there are only stars up there.

We have to accept that this has been the norm in all history---one man's best friend is another man's best meat, even though we're all human. The religions lost their power of human control when people wisely abandoned faith in a common destiny, because they realised that it was a mystical hoax manufactured by their fellow men in no way holier than themselves. There is a good lesson in this for all of us, but human beings are not made for taking lessons from their fellow men, that is why the religions failed---ie, the very adorable, critical faculty that makes us what we are is dead against human manipulation. All this became standard with the rise of science; for the rise of science is the application of logic (instead of mere opinions) to human life, as Russell showed in his History of Western Philosophy.

Yet still, as knowledge, all that is sheer mystery, or computerised mystery: that these things just happen to come to exist. They send signals around; our brains capture some of these signals and computerise knowledge out of them. But the greatest wonder is not the vastness of the cosmos, not even the stupendous sizes of some stars of which, they say, about a hundred million of our sun can find room in them, but the brain of mankind which discovers all that through knowledge calculated from the signals of objects only to die eventually, decompose, and disappear. That is the story of life. The question is, why can this tiny brain do that? I envy those who claim to have the answer either religious or scientific as Russell supposed.

However, all that has nothing to do with the logic of time in the universe, as I call it. Time can be interpreted in many different ways, so it was easy to wrap it up in mythologies and religion (being imaginary suppositions), but there can be only one logically approved time in the world. And we cannot say time in that logical sense does not exist because there are always human activities which require or create time sequences[16]; all the mathematicians had to do was to invent a mechanism for expressing 'intervals between points'---that is all time means, a sense of waiting between events or points which, once mechanised, can be used to time all other events; thus we have regular time clocks no more mysterious than language and speech, and based on nothing more than the intervals between points, activities or events---ie, purely secular. The promotion of legends and endless mysteries of time, on the other hand, is the pastime of the religions not scholars of rational thought, philosophers and logicians. Of course, in society time is seen by people in a thousand different ways. That is forgivable since

---

[16] Talking about building on the shoulders of giants, great thinkers like Bertrand Russell and A.N. Whitehead would not have defined time and time sequences if they did not exist in logic, with Russell asserting that there is no longer a universal time, etc. For this is a serious debate, as it goes to the roots of human existence, action and destiny---or no destiny.

time is rather difficult to explain----I use more than ten realms of A-4 papers to write one book on time. The problem is that the logicians and scientists hold our lives, even the safety of the earth, in their hands; and so it is they we have to listen to not the servants of God, or the peddlers of mythologies who outnumber sane and rational men.

However, in this sense, as time is defined above, or in visual and realistic terms, time is immaterial; it is only a concept we apply to nature for human guidance. The only way to trace time physically is in growth, particularly the growth of vegetation, since human growth is rather subtle and invisible.[17] Hence the last minute is nowhere; it's been used to live imperceptibly. We only have had a brief history of existence. It is the existence itself we want to question; we want to know what for? it is obviously silly to say 'stop the world I want to get off'---to go where? Yet it's got a point. Time is not to blame, and it does not go anywhere or cause anything to be---it's all chemistry and natural processing. Time is absorbed and used to live. So the usage of time is the passage of time, or the passage of time is the usage of time. Understood that time is not passing through the universe like a thread, and that even the cosmos has no time at all since it has not yet invented the point and acts only through random accidents. Now let us consider the mysteries that confuse our scientists, some of which are due to the wrong understanding of the nature of time, for it certainly is not just there, and to say it does not exist is not logically tenable---we go to work by time, for instance: we also travel by time, sleep by time, do sports by time, etc.

---

[17] In many PhD researches in biology students can record the time developments of some plants, but even then only in days or hours.

Gravity and Time: if time (as 'being-in-miniature' created with mathematics in the mind only to guide human actions)[18] is not passing through nature physically like a thread or stream then it cannot be affected by anything, rain, storm or sunshine, except by the motions of the earth, because, of course, it is not physical or touchable; If physicists insist that it happens then the Einstein explanation of the Twins' Paradox comes into play, namely one of the clocks involved in the experiment may have experienced acceleration. Or there is no shame in saying we just do not know what it is. After all we don't know why we are here or where we came from and have been dying in millions without any recompense. The same response can be given for the Entropy and Time query, except that here I am confused and annoyed by the misuse of brain power or lack of it. Things happen in the cosmos without the dictates of time, or coincidental with time because there is always time. Why should we suppose that entropy alone is caused by time running all through the universe, chasing entropy to make it happen----especially since we know that our discrete time cannot do so? After all, existence or history is not the march of time but the march of events. I agree that some activities may betray the signs of time lapse because duration can result from many kinds of events which can be long-lasting. Alternatively we can say we simply do not know; or maybe they accidently happen every thousand years or so. For we know for

---

[18] You look at daylight and say it's moving by seconds towards darkness but in your mind only. In nature or astronomy there are no nights and darkness. But this brings in the question whether there are several facets of reality, since we are also the products of the same astronomy. This is a metaphysical conundrum much exploited by the religions in such saying as 'We are all the children of God', the good the bad and the ugly---are we? These are some of the questions that make me think that Sigmund Freud was one of the greatest thinkers of all time as stated below in the text.

sure that time is not running through the cosmos to go and cause entropy somewhere.

Most of these 'scientific mysteries' derive from the unexpected, 'electric' shock scientists received from the miracle of time dilation. But scientists forget that the dilation of time is known as 'The dilation of time as a measure of moving clocks', so Einstein's interpretation of the conundrum that it occurs in a different frame is sensible. My own suggestion is that clocks do not control time, and clocks can mal-function in all sorts of ways. There is so much we don't know, can't even begin to realise exist. So the trouble with clocks can be called 'clock dilation' or a mechanical curiosity---for maybe the clocks involved in the experiment had experienced acceleration. There is no need to jump to any imaginable (and probably religious) interpretations. What has been physically proved is that there is no universal time, and that the time we have is discrete, being the product of points as applied to space. It cannot flow through the world or fly through the air. The movements we notice are the motions of the bodies which produce the time lapses or what we call 'time', not the movements of the time itself. That is all we know and can prove in logic, otherwise we can confess that we simply do not know. There is no shame in that. Researchers are being unduly influenced by science-fiction writers, who have proved nothing, except that they have fallen in love with the notion of time running slowly with speed; and they stress that it would be useful in astronomical journeys to other planets. Yet we do not build scientific theories out of human desires. The point is time is not universal, it is not running all through the universe, and it is in fact discrete because it is created with points out of the orbits of the sun---a mathematical entity and strictly discrete---based on the space traversed round the sun as explained below.

There are many things in life we don't understand. One of them is Albert Einstein's promotion of the genius of Minkowski when the latter had clearly distorted relativity. I suspect it is because he was glad that the mathematical equation of space to time (though logically flawed) helped him to dispose of (or ignore for the time being) the many problems facing special relativity and time. Einstein had made time secular, as our own local time. He said he got the idea from Lorentz, without solving the basic question "What is Time?" Yet Bertrand Russell was asking the question, "What is measured by the clock, if cosmic time is abandoned?" It was like this: in his book Relativity (Part 1 -17) Einstein had stated the Minkowski formula and asserted, "We must replace the usual time coordinate t by an imaginary magnitude √-1.ct..." thus, apparently, equating space to time by means of mathematics, as the conclusion is $s^2 = c^2 t^2$, clearly aimed to sweep away many problems including that of how the muon manages to reach the earth, etc. The problem is that the equation was logically flawed---ie, if i is used to replace t, then what does the ct stand for? And if the basic logic, the premise, is flawed then the whole equation is false, and therefore space and time are independent of each other, except that time is derived from points, which requires the use of space. Therefore time is 'space-time' in that sense.

So the problem of time's essence has come back to haunt us, as I made clear in my book The Coming Revolution in Physics (never published except by myself and totally ignored.) Yet the problem is there. As stated by Einstein himself, the formula given above is supposed to be the mathematics for launching 4-D geometry as Minkowski proposed. Needless to say, it has entered physics with such far-reaching implications that we may never be able to expunge it completely to the end of life. Yet can anybody seriously suppose that the above formula, as logically defective as it is, is a true statement of reality upon which to base the whole of theoretical physics? No wonder we are bombarded with so many fanciful theories, of worms, multiple worms, and increasing dimensions of reality so much that physics looks like an Easter Jumble Sales Market on Clapham Common. And I've been saying such things for more than twenty years. They will not even honour me with an acknowledgement of my communications.

However, what the Norwegians have discovered about time is interesting, although what they infer from it is not correct. Also, the Press report mentioned time as 'a subjective quirk'. I use the phrase 'planetary quirk'. In all my books I argue that whatever happens to us on our planet is a planetary quirk that has nothing to do with the universe, and that we are on our own, and reality ought to be re-defined. There is a book waiting to be written, called 'Reality in human perspectives'. For the universe is so vast that objects as big as the earth might only register as some kind of a minute grain of sand, let alone the still tiny objects on its surface. (See the Times of London, 19-6-19, pp32-33.) They have evidence that there is only one day, yet I have been saying this for years, so I can confirm that what the Norwegians have found is true, but it also means that time cannot be denied---as they are saying---because obviously we have time, but it is compounded of some features of the world by man and not bestowed by God. The elements for creating time are there in nature, but it takes the human mind to put them together. In the words of Bertrand Russell, as I have stressed all through this book, time is constructed, and Einstein also told us that it is limited to a frame or planet and that there are as many times as there are planets. Whether the Norwegians got their idea from some of my books or not, they've certainly made my job of persuading the sceptical world about the theory of secular time much easier.

But objects do not contain time merely because they occupy space, for the attempted equation of space to time was not successful. They have covered this up, but it is the truth. Time is the product of space, not the same thing as the space. According to Bertrand Russell time is constructed by man; the materials for the construction may be scattered all over the universe, but it takes the human brain to put them together. Therefore, as always, Einstein was right; time is secular, but it means there are lots and lots of things we do not as yet understand and must keep on learning. Scholarship is the creator of inventions.

To conclude this Preface, let me stress that our SI of time is derived from a combination of mathematics and metric space; they create for us moments of existence, as recorded by the clocks. When time is regarded as 'just is' the other units of time are also taken as 'just there' for us to use. Yet this was wrong and has perpetuated the mystery of time in which the religions take great delight. In fact, secular time (which really began with Einstein and Bertrand Russell), imply that there is only one second. **The whole of time is reduced to one second as our SI of time. The other units are the multiples or fractions of the second. As the second increases in numbers to become two, three, four, ten and so on; or in minutes, hours and days, time is going. It is another way to account for the passage of time: time is known only in units, and as the units (like the years) increase in numbers, time is going or passing by--- it is also the same thing we use to age---hence, very simply the usage of time is the passage of time.**

For more clarity, let me stress again that time is the combination of mathematics and metric space, creating or acquiring specific amounts of existence for the earth---we can't have any space for existence until the earth has moved a certain distance that is covered and bountifully supplied by the sun; and that space is one moment of existence to mankind although it is miles in space. That is what the SI of time means. It did not drop from the sky. So for the other units of time we simply multiply the space or duration of the second. Thus we do not convert space to time; we create time (we purchase the duration with space and this was all done with mathematics---for free). As already mentioned, you cannot have intervals without points; and you cannot have time without intervals; both are human in origin. Life on earth could not flourish without them. It's the old story of Creation with the mathematician as God.

We create time with space and Bertrand Russell was right when he deduced that time is 'a construction' due to the facts stated above. Only a great mathematician could have known that. The irony is that the mathematics is the same whether we call time divine or secular. It is not new, only the logic has changed. The reality remains the same. The only advantage, in my humble opinion, is the rational concept of the passage of time as the same as the usage of time. This is where the revolution takes off, for time consisting of multiples of seconds has to be necessarily discrete; but discrete time cannot march through the cosmos; discrete time is spent unit by unit---so history is the march of events not the march of time, for instance. But if time cannot or does not march through the cosmos then where are we, or what are we?

Yet if the reader goes to Professor Arthur Eddington, he or she will find him furious that some people still think that time is flowing through the universe (The Mathematical Theory of Relativity Ch.1.1.) So can we think the unthinkable to round off---that many or most of the problems in science about time are due to the misunderstanding of the true nature of time as mentioned above without having to go through all of them one by one? They have ignored me for more than fifty years, but the problem is still there!

In sum, time is known in units due to the manner in which it is physically constructed or created, as evidenced by the SI of time. Thus time and timing looks like counting cycles, so many kilometres equals one second on earth cyclically, making it seem as if we buy units of duration from the space traversed by the earth round the sun.

INTRODUCTION

Time is human nature; all creation by time is also human nature, including some animals (probably all creatures), meaning it is their natural way of acting or living. For even walking takes time, but we do not worry about that. To mechanise this time in a clock as the motivator for planning and futuristic action is the bone of contention. We have the same problem with speech and language, literature and thought: to make verbal noises is natural; babies do it all the time. To create language and literature out of these verbal noises is another matter.

In the absence of sentient beings there is no time, instead all action is randomly caused either through accidents or chemistry without time and timing. For time requires points and so mathematics had to be invented before time reckoning could be contemplated.[20] Otherwise the risk is to misunderstand the world since without sentient beings time cannot exist. The religions made that mistake; science and rational thought must not

---

[19] Time is a difficult subject; in fact, the most difficult subject in our lives, and so these quirky devices in my presentation need excuses from the reader, after all what we all want is a clear idea of what it is and written out as lucidly as possible.

[20] Mathematics is not only important as the basis of rational thought; it is also partially the basis of being, or continuous existence due to its control of time, since time, also, controls everything we do. The reason it was previously regarded as divine. I believe a deeper understanding of these issues will show that the inventors of mathematics---most of whom were religious---were the greatest thinkers since in theory all this is implied in religious thought without a shred of evidence that God exist, just playing with the notion of 'Father Figure'. I give them credit for knowing how to manipulate the human mind most effectively and all politicians know this too.

repeat it. Time lives in a brain, conditions it, and takes over---therefore its origin must be activity, most of all, 'planned or consecutive activity'. There is no time outside man and animals. Even the universe has no idea of time because (and I repeat) time requires consecutive points created by the brain. Hence saying "The usage of time is the passage of time" in what follows is to suppose that, logically, time and existence are merged in the human mind. There can be no independent existence of either, for even moving a finger is 'an event between points'. It takes time to do so. Thus I define time as **life plus activity**, *and since the activity comes after the life, all time (the theoretical essence of time) is inferred from historical events for application to current and future events. And it grew minutely to reach its present perfection. In the absence of a clock, we betray this notion of time in statements such as this: "When the shadow (of something) is at such-and-such position, it means the time is so-and-so and the dinner must be prepared, etc..."*

Time is therefore more mysterious than life, and, being historically rooted, we can't even live our lives without it. The religions exploit this by telling us we are made by time and that it is the begetter of all existence; that is why a book called A Brief History of Time can be written and be so successful by one of the clever begetters of human deception. All religion is based on time, and so closely associated with life that it is virtually the same as existence. Yet, to be truthful, time's true essence is revealed by the fact that it requires points----man had to live for centuries to know how to invent the point, therefore time is human in origin. In the absence of such a convoluted explanation of time, the metaphysical nature of time can never be known, since we never know what it is we're defining even when we use high-level mathematics, and saying it just happens to be there---as 'just is'---is admission of ignorance.

The reason life and time are bound up together is that we never can have existence without events since the intervals between events are what we know as time, or use to construct time sequences. But to use activity for constructing time sequences it must be recurrent or planned, as a solitary action cannot give you the minutes passing by. Without activity there is neither the need for time nor could we even have the sense of time---as the **timing** *of activities. To understand this is to know the essence of time. It is based on structured, planned or designed activities and so sentience is required. There will be time wherever there is sentience.[21] It is the reason the universe itself can have no time of its own—ie, without consciousness, for consciousness must assign the points for the repetitive cycles (of events) that can be used to set our time sequences; precisely as Professor Richard Feynman put it "...something that happens over and over again..." that we can count as the rate of the passage of time or, in short, use to reckon time: I repeat, without points there can be no intervals, and without regular or repetitive intervals there is no time and all events occur through random action. It means our time is unique as it has a dose of our nature in it.[22] So is our technology, probably our science too. However, even though*

---

[21] **It now appears that the universe itself is senseless, but generates trillions upon trillions of objects and chemistry that some of them accidentally result in sentience but only temporarily, and one of them is mankind. The problem is how to define sentience. I think it is the rare ability of some objects to use the signals from other objects to regulate their essence or lives electronically or chemically, but not through the quantum except in conjunction with the electron as per QED. True or false this may be called the definition of intelligence in scientific language---under duress!**

[22] To deny this amounts to saying our nature is not part of the stuff of the universe. I must say many assertions imply this idea, but they're all illogical. We belong somewhere. At least I know where I belong to---up there among the saints!

these two essential elements of existence (consciousness and time[23]) are apparently merged, it does not mean they constitute one entity that might have been created at the same time, because time requires points that have to be supplied by existence, or the 'Being'.[24]

As a matter of speculation, I think intelligence derives from judgement; judgement from comparisons; and comparisons from multiple-perceptions, mostly through the data of sight. All of these can be created by criss-cross signals between objects without any directions from a sentient being but rather capable of generating sentience in certain forms of matter. The silicon, for instance, is recognised as an efficient transmitter of signals, and behaves like some kind of a crude mind. This is what I personally can think of as the rational basis of the human mind, especially as simulated by the computer. Of course this may be mere speculation, but it helps us to imagine life as something that came out of lifeless matter by being built up bit by bit around a core, created by signal transmitters that had cobbled together through accidental collocation. The idea may be pushing logic to the limit, yet it makes death look like disintegration rather than destruction---the disintegration of the original collocation rather than the destruction of matter, since matter is never destroyed; it is always a transformation from one state

---

[23] I think the secret of life will be found in the connection between consciousness and time.

[24] Einstein was honoured with the grand title 'Philosopher/Scientist' mainly due to his theory of gravity; but to me it is for his theory of time, and the crucial idea that there are as many times as there are bodies. He is the greatest thinker of all time because he abolished absolute time to confound the religions. For time is the most important thing in life, as the life cannot even be endured without time. Close study of time reveals metaphysical ideas that go deeper than that of any other subject----but what is it? Lazy philosophers and scientists regard it as just is; but even so as what? How can we define it logically?

to another. So the answer to the question, where do we go when we die, may be that we return to the original matter whose collocation created life---earth to earth, dust to dust. The crafty religions always have the last word. They are very good thinkers whose only crime is the disregard of logic and science.

In fact, time, speech and language are the real mysteries in the universe as stated on the first page of this book. Life is rather logically straightforward and inevitable, bound to happen over and over again. What makes life important is that without life there can be no speech, language nor time, running through a central transmitter of signals (like the microchip) which we experience as "consciousness".[25] But philosophers have discussed God for centuries, yet in what language is he supposed to communicate with us? Besides, God is not the only controller of human events, as we do not know what role dark matter plays in all this---ie, in the process of life, its transformations and disintegrations. Perhaps even death is not what it seems. It might be easier to bear, barring economic and social implications, because there is no such thing as the total destruction of anything in the universe; so it may well be that death is transformation of life into other forms of matter without the conscious memory of the past, due to the disintegration of the central, signal control system. Again the religions have the right language for this, saying that the dead Jesus is always with us----all those we can remember too, I think.

---

[25] A conscious person is scarcely different from a robot, except in the efficiency of his central signal control or his brain, which is billions of times more complex and therefore more efficient. But we get to know how the brain works from the nature of these robots.

The Origin of Secular Time

One implication is that man created time by using elements (signals) found naturally in nature. And that is the closest we can get to identifying the essence of life and we owe it to Albert Einstein, H.A. Lorentz, Bertrand Russell, Professor A. N. Whitehead and Professor Sir Stanley Eddington together with their followers, the present writer counting himself illegally among them. Let us note, however, that Einstein made Russell a rational philosopher; before that he was a student of Hegel---and it's rational philosophy that did it. This writer is already celebrating the triumph of rational philosophy as against all other lines of thought. But the tendency of mathematicians to imply in their conclusions that God Is, causes irritation and misinformation. For instance, there are no such principles (as Roger Penrose wrote about in The Road to Reality), that we are capable of finding them, unless religion is implied; yet that is a philosophical idea but religion and philosophy are not bedfellows; in fact, for the rational philosophers, since Einstein, they've been diametrically opposed, since, by physics, the clear evidence is that the universe has no time sequences of its own, no planned existence, and no principles----without God. All action seem to be caused by random events, one after another. As such, there seems to be no evidence of the laws of nature, except that some events and their effects seem to last forever, long enough in human terms for civilizations to rise and fall before fading; many of the events we associate with, or attribute to, the laws of nature fall into this class. Only the mathematicians believe otherwise.

Yet if science is anything at all in epistemology (as we are forced to believe because of the Atomic Bomb principle that if you ignore it you're dead), then everything in nature is caused through random events, not Penrose's "...underlying Principles that govern the behaviour of our universe..." (Page xv, Preface) That's arrogance. Man knows nothing of the kind. There are no such principles; all the evidence points to random action without planning, regulation or destiny---after all, either the Steady State or the Big Bang theory revolves around recurrent acts of material evolutions. Mathematicians ten to imply religion subtly in their conclusions; on the other hand, philosophers accept that man is worthless; the planet we live on itself is worthless and so insignificant in the universe that even if there is God he must have better things to do than look after us out of these billions of mighty stars, especially when it comes to individual persons. Otherwise who decreed the 'underlying principles'----and how deeply underlying are they? How do ordinary educated people who are not Oxford professors of mathematics apprehend and obey them? Do they include the Atomic Bomb, and if so why, who planned the whole terrible thing and therefore what is the purpose of life? Many scientists believe that philosophy is easy or unnecessary. In fact, the German Idealists and Wittgenstein made it look as if we can dispense with physics, but since Einstein the rational thinkers have warned us to obey physics or die, and I doubt there is any human being who does not agree with that warning. However there cannot be a better ground for amicability not hostility between the post-relativity rational thinkers and scientists. My other concern is what we do to their books when they die. Surely we can afford to make their books available at all times, eh? I am thinking of scholars like Russell, Eddington, Sir James Jeans, Einstein, Whitehead---the list is long, yet we can afford it and the benefits will more that pay for the costs.

First, the philosophy[26]: the ultimate idea upon which this interpretation of time is based is that all time is known and used in units only and the word 'time' is culturally meaningless unless it is thus quantified. Silent time is mere existence without activity because time means life plus activity. Therefore the old Leibniz theory that time is 'a succession' is correct. What was missing is the notion that time exists in units, and in units only; it has to be in units to advance or occur in succession. That is what Leibniz didn't know. But it doesn't seem likely that even if he had known it the religions would have accepted it as true to undermine the concept of absolute, cosmic or divine time.

We all know that religion is closely associated with time as they are all based on it. So, right at the beginning, let me state my belief for the benefit of the reader: religion may be a load of crap to some people, yet other people too think their lives have no meaning unless their religion sanctifies their very existence. You can describe such people as intellectual morons, but so long as they have the right to live, they must be given the right in law to keep their religion. What is obvious is that the religious theory of time is so wrong that it makes the entire religious view of reality seem to be a joke, especially the story of Christ[27]; but religious persecution is a gross act of philosophical ignorance unworthy of Homo sapiens. If by some miracle we could all convert to the Theory of Evolution the world would be a better place. It would be a religion

---

[26] At all times and in all circumstances philosophy is indispensable because whatever common sense and science discover or find to be there in nature, it is our philosophy that can place it in human terms and ask the relevant questions to clarify for human tolerance. When it is upgraded to the philosophy of science then man is placed in the lofty, safe hands of thinkers like Professor Eddington, Professor Whitehead, Einstein, Bertrand Russell, etc. In this sense even religion has to have a philosophy upon which it is based, to tell man what is good in its sermons. A rigorous examination of this nature will help weed out all those crooks and their phoney religions.

[27] "Justin Welby, the Archbishop of Canterbury, has appointed as his 'ambassador' to the Vatican a priest who denies the physical resurrection of Jesus"---The Sunday Times, 13/1/19, p7.

with nothing to worship but the grace of mankind not the Grace of anybody's God---and wouldn't that be better than the cacophony of sectarian madness and brutalities we have around the world at the moment?

However, strictly dealing with time (and religion is mentioned because the believers always have a say!), in rational or logical thought, we simply have to agree that to get the SI of time the earth cycle has to be divided into specific units of duration, the ultimate part of which is the second. This has made it possible to align or link life, time, and conditions on earth to the earth's journey round the sun for time to guide activities on earth; thence units of time have cultural (even scientific) meanings.[28] And all units of time are either multiples or fractions of the second; when local time was discovered this became the scientific basis of time or secular time---therefore all time is known and used only in units, including silent time or sleep-time, because in all cases the exactitude of time can only be stated in units, otherwise it would not be understood as the expression of time or earth time. I have to stress that without mathematics none of this could have happened.

Also if the Leibniz notion had been well analysed, secular time could have been known long before Albert Einstein was born. It is relatively easy to explain how a succession of time units pass-by through usage as they are absorbed and used to live one by one. For example, where did the last second, minute or hour go and how? The answer, as I argue below, is into the growth of objects and people. Age is the accumulation

---

[28] We now call the meaning of time cultural, but in time past it was called divine. When local time was discovered divine time ceased to have any meaning that is why we call time 'cultural' or the product of our culture, science, logic and rational thought.

of time units, like the years. Therefore the independent, physical passage of time is a myth. Where from and where to? All the grand theories of mathematicians and bishops about the passage and eventual destination of time are the products of ignorance disguised as knowledge from high. You could say that they were trying; but put forward as suggestions no harm is done, except that they never had the humility to argue decently by logic---which is very annoying.

Time is nothing but life plus something else, because you have to be alive to know about time[29]; whether 'that something else' is divine or secular is the question, but it makes time discrete, with all the momentous consequences detailed in the text.[30] This amounts to giving a clue to scholars as to how the theory in this book evolved, or came to be conceived once the mythical religious something is rejected. Above all, we could never have had the all-embracing time system without mathematics, which came late, not in time, but in human existence. It is wrong to hang everything in history on the march of time through nature as there is no such thing.

However, to get on with the narrative, now, as a result (and rather late), this discourse is in the footsteps of the relativity definition of time, and in that regard, it is assumed as self-evident that metaphysically time is the most basic thing in life with which everybody is familiar as being as strange and unknowable as life itself, the reason it was once thought to be divine; but in this monograph time is discussed without regard to how it has traditionally been known in theology, moral and pastoral theology, (the nature of which I don't know),[31] science, philosophy, and

---

[29] Also, like language and speech, time and the sense of time have to be learnt; it's not in-born like blood circulation and sound.
[30] To my way of thinking, time is life plus activity.

also by the ordinary man and woman, the sort of debate that gives philosophy a bad name. Rather time is presented in a logical argument that is all based on Bertrand Russell's question: in the absence of a universal or cosmic time, 'what is measured by the clock?' I am hoping to convince the reader of the true nature of what 'it is', consistent with man's current means for attaining true knowledge of the world, which may be called 'post-relativity physics and QED' by whose rules we have to re-interpret everything by the quantum or the particle of light.[32] Living in a world that has been turned upside down by the quantum theory, philosophy cannot ignore it and continue to debate the garbage in some strange thinkers' thoughts. That no thinker, scientist or philosopher, could ask such a question about time before relativity, shows the depth, breadth and momentum of the revolution introduced by Albert Einstein.

---

[31] Human beings have invented a host of bogeys with which they delude, entertain and frighten themselves, crush, and eliminate, even murder their enemies and opponents---either individually or en masse as in wars---and in the end go home meekly to worship totems like morons to satisfy a queer desire to humble themselves before their own bogeys---and they say we should regard that as the meaning and purpose of life. I beg to differ. To me it is the misuse of brain power; more intelligence is expended in acts of war than the search for the good life for mankind.

[32] Of all the billions of things and concepts in the universe, QED is the most important and mysterious, because the quantum is what we see as light, the light by which we see the world and all of reality; and light conveys the images of things to our eyes 'after a time lapse'; but this time is man-made. How can the mad ideas of the German Idealists have something interesting to say about this? The point is that we know all this because of Einstein. So there ought to be 'Life before Einstein' and 'Life after Einstein' intellectual movements for thinkers to know where they belong and for what use. Apart from the classics, almost all philosophy before Einstein will become redundant, including German Idealism.

47

The Origin of Secular Time

Unsurprisingly, therefore, time appears here as completely different from what the ordinary person calls 'time'. In other words, this is philosophy presented strictly as a logical discourse (as Bertrand Russell stressed all his life in numerous books)[33], not as the subject has traditionally been debated in philosophy or theology, since the aim is to contribute to the solution of the 'new' problem of time---how we get it if it's not cosmic, and how it passes through nature 'with a history and a future of its own', if it is merely conceptual[34]---and certainly not for the glorification of philosophical traditions, the cantankerous German Idealism (a branch of linguistic philosophy now regarded as sheer Teutonic humbug) and Wittgenstein, with whom philosophical institutions have fallen in love against Russell's advice.[35]

---

[33] This is adding to the call to do philosophy as he told us.

[34] Time has never been seen as physical in anybody's theory; writers just thought it is just there---'just is'. Only Leibniz came close, saying it is a succession in his 'Letters'. Now we know it is a succession of time units, like the year, pared down to the seconds. Actually time is not physical but conceptual, a combination of psychology and mathematics and therefore purely human in origin. As such it cannot be part of physics; if we find that it is needed in physics then the supposition that man makes his world is true, and history is the march of events not time, while the future does not exist and therefore unknown. "Time and Physics", or "Time in Physics" is destined to become a long debate in physics and philosophy.

[35] It is conveniently forgotten that Bertrand Russell and Sir Karl Popper condemned philosophy from the time of Kant to the 20th Century; mainly in Russell's History of Western Philosophy, but, in fact, Popper said just before his death that he was ashamed to be called 'philosopher', due to current fashions in the subject---Wittgenstein, Existentialism, etc., for how can we destroy physics? "Russell saw these things in that light, and so did I", he said---see Modern British Philosophy, by Bryan Magee, Secker & Warburg, London, 1971. Habitually human beings turn their backs on their heroes, what is said today is forgotten tomorrow, that is why we have recurrent problems.

Samuel K. K. Blankson

Philosophy has never been a strict academic discipline; the few areas now presented as relating to physics and astronomy, psychology and chemistry, came about in the last two centuries, mainly due to the works of writers like Darwin, James Clerk Maxwell, Russell, Planck, Eddington, Einstein, Whitehead and a thousand others encouraged by Darwin. Otherwise, on the whole, philosophers teach the personal ideas of some thinkers and the professors choose their own heroes where one man's meat is another man's poison, for even the religions have their own philosophers, sometimes known as 'gurus', and suspected of frauds, financial and sexual offences. This writer is having none of that due to the ideas of Russell and Einstein, meaning their new ideas in logic and physics. Readers who think otherwise will find a welcome home in the intellectual jungle mentioned above.

For centuries Philosophy used to be idiosyncratic but Russell changed all that; also physics used to be inconsistent but with Einstein and QED it is no longer so backward and, thanks to the Planck Law, it is, above all, no longer tentative. We do not only know the truth about physics; we even know how to use it to destroy the earth, our home. The reason Russell urged us to study the philosophy of science, which is why he eventually rejected Wittgenstein. Nobody should blame me for my harsh views about how Wittgenstein is worshipped in Oxbridge.[36] I have very good

---

[36] Of course Oxbridge is great not because the British boast that they are good but because of the intellectual work they do. Oxford has my highest marks for OED, one of the finest intellectual achievements on this planet. And Cambridge too has the Cavendish Laboratory as well as being the institution where Newton and Russell did some of their noble works. The demerits are just as momentous---they promote religion far too much. The sort of people who threw Russell out of America and could have done the same to Einstein before he was famous, in the name of God. Thus the Indian 'Guru', Srinivasa Ramanujan did some good work in mathematics at Cambridge but he also said, "An equation for me has no

reasons for doing so. We have no choice but to respect the forces of nature as revealed in physics; we can never stand against them. A Philosophy against physics is the philosophy of death, and Russell stressed this in many books and lectures---"Remember humanity and forget the rest", almost his last words. We live in the world of ideas, made what it is with ideas; the ideas from teachers; the teachers from theorists, and the theorists from philosophers whose lofty thoughts come solely from the logic of human existence and the earth is our only home. No kind of writer, called 'philosopher' or not, is going to help destroy this home so perilously balanced within the laws of physics--- however those laws were made is immaterial.

Time is second in importance only to life itself; but altogether I believe I have advanced sufficient arguments to support these conclusions, bearing in mind that these are one man's fallible conclusions: (1) there are no days in nature, there is only one day. (2) The rotating nights are astronomically irrelevant, even a nuisance. (3) The religious 'day of judgement' is meaningless because there is only one day in all nature--- the sun is on all the time. Days of the weeks, months and years are all human concepts of no relevance in the universe; we often forget that we germinate variously like all other creatures on the planet, and, as creatures of the planet, among other billions of objects, we simply do not count. With all our gods we are still worthless in the universe. Can anybody tell us how our gods will fare when we're all dead? Being clever is not enough. Colonising other planets is futile; there is no eternity anywhere. The old idea of eternity evaporated when local time was discovered. Secular time has opened the door to the most far-reaching concept of humanity that defines man and his future in the most painful

---

meaning unless it expresses a thought of God", which is nauseating.

and unflattering scenario. Even science can't do a lot about that. In the end we'll all die; deaths and renewals even among the mighty stars are commonplace. Stupid or clever what is the purpose of life and what will happen to nature when we're all dead? It's no longer a question of 'when time comes to an end'---that supposition is gone forever----'but when we're all dead'. A clever brain just because it can ask such questions is no help if it cannot provide the answers; so in the end our brains are the cause of our woes. I am praying to come back after death as an idiot.[37] (4) The origin and nature of time is now known to be secular. (5) Indeed it is regarded as humbug to think of a universal time to cover so huge and complex a universe all at once as fixed and absolute. (6) Being constructed out of space with points, time can only be discrete---that is, known and used only in units. (7) Logically time is based on 'being' not being on time. For time reduces reality to periodic chunks; another expression is that time is a mathematical device for reducing reality to shorter periods---hence the whole of DAYLIGHT consists of twelve hours as shorter periods, and so forth. And the hours can also be broken down to other sub-units of time as we find culturally convenient. We can have multiples of seconds in this daylight up to no limit (eternity) because it is permanent and always in motion. The time itself is not in motion; it is rather applied in numbers to the moving reality as multiples or fractions of the second; for time does not move. (8) Time is life plus activity, therefore they are virtually inseparable but it

---

[37] By the way, as regards the efficiency of the brain, I take the view that every object has built-in mechanisms for defences and survival; the brain's mechanisms only happen to be more efficient as it has to serve trillions of cells (a huge volume of work than any other organ), and that it is the efficiency of the brain in carrying out this function that seems mysterious to us. It also imply that mistakes will occur in such complex activities, hence the problems we are all too aware of regarding illness and accidents, alcoholism, drugs, etc.

makes time secular. Existence plus activity means but for activity there would be no need or sense of time and timing. We have proof of this from the newly-born. They do nothing and have no sense of time. It would normally be years before they learn to read the time. (9) Secular time based on points and known only in units can only be discrete. (10) The passage of discrete time can, also, only occur unit by unit, making cosmic time and time travel untenable. (11) Therefore history is not the march of time but of events. Of course, the two are often conflated through ignorance; but things and events move on physically; time can only advance mathematically through numbers, number of seconds, minutes, hours et al, that are quoted in association with events and the positions of objects.[38] Hence the Leibniz concept of succession is relevant, an important concept so long ignored. (12) Space and time remain separate entities despite the Minkowski proposal, even though mathematicians still believe in 4-D geometry based on his defective 'ict' equation. (13) According to Einstein, the discovery of local time means time can begin from any space; I add that sentience and arithmetic are

---

[38] Thus we say 'John came at 6. Pm'. The coming is subject to physical motion; the time is mental and subject to mental operations only. So the time is variable without any outward changes, and instead of 6.pm, it might be 7, 8, 9, or even 10.pm---that is purely numerical, mental or arithmetical, not physical; the same way time passes by, namely in numerical units not physically (for we count cycles elsewhere as units of time and apply them to reality as 'time'.) But the physical and mental aspects of reality are traditionally quoted together as a matter of logic, realism and social practice; for another of the mysteries of time is that once invented it became inseparable from life. The mind and knowledge of time must have evolved together. What makes us human must have grown up together with the sense of time and timing. This may be provable through fossil records, or perhaps it's too late. Activities of chimps indicating that their minds are growing betray knowledge of timing. Indeed the sense of time and timing may very well be the foundation of the mind and thought---we can see the chimps gradually acquiring the ability to think and timing is an integral part of it.

required, making it completely human in origin. (14) Time is unstoppable because it is based on the orbit of the sun by the earth--- which nobody can ever stop---and therefore about 31, 536,000 seconds equal to one complete orbit, and we start another orbit, or another year: our principal unit of time from which all other units are fractions, as well as being multiples or sub-units of the second. That is the unique nature of time. Each unit is a fraction of the year and at the same time either a sub-unit of the second or multiples of it. Sub-units of the second are well-known in science. Multiple units of the second are a reference to all the other units of time, including the earth-year and which are fractions of the same earth-year. That is what logical analysis of time reveals, without which time could never be understood in logic as opposed to religion. The symmetry was absolutely vital---one of those metaphysical truths, like evolution or the curvature of space causing gravity, upon which the scientific culture is based.

In reading any philosopher one expects something about epistemology, or how we know things, and over the centuries it has been decided that our intellectual ability is based on points and instants.[39] I think it should now be points and intervals, not instants. The instants are value judgements.[40] Points and intervals can be seen as beginning even from

---

[39] Instants were supposed to be 'given' as divine or cosmic time. Under secular time it is the same scenario, but 'created' by man, not given by God---because we can create local time, our own time. Gaps between cycles are what we use to create units of time, the most basic of which is an instant or a moment. All of us on this planet use the earth-year and sub-divide it down to all the other units of time that is why we do not physically count cycles to obtain our time units. However, in mathematics or metaphysics every unit of time is part of the yearly cycles, and so it means we still count cycles to obtain intervals as units of time--- from instants to infinity/eternity or even down to fractions of a second. Again and again I have to remind the reader that the creator of local time failed to notice its significance. It was Einstein who said it is 'time pure and simple.'

the womb since the unborn are known to kick about (not only for fun but also as they turn and turn). Eventually this would become the basis of our intellectual abilities (points and intervals in succession built into the embryonic brain).[41] Language is not yet accounted for, but language is based on sound and imagery; both of which could develop from contacts in the womb---it is known that the foetus can react to sound, to be linked with the data of sight later on in life for language development. Even where there are no obvious signs invisible psychological changes might occur. This process is speeded up after birth; hence language is quickly learnt by infants where there are no biological defects. The mind easily matures through education.

I have no doubt that some readers will argue that the Introduction is rather long for such a book; if they would only say it and not abuse me on Social Media, I would regard it as an act of kindness, to make me think that the Social Media could somehow redeem itself away from the lunatics who are giving it a bad name. Nevertheless, let me explain that time is the most important subject in all literature and our intellectual

---

[40] These value judgements are mere assumptions. There are two theories of explanation to account for how we get time in units. The divine or cosmic time theory is that time just happens to be in existence, that it just is there, and we use God-given mathematics to sub-divide it into the various units socially convenient for us, or our style of life. On the other hand, the secular, logical and scientific explanation actually shows how the units of time are created as fractions of the year, so that a specific amount of seconds equal to one full orbit of the year, called one year---and what is more important, that point is where we start to calculate another year. In theory, time units are derived with points from the orbit of the sun as relation between points, such that it is easy to follow the process physically without any appeals to mythologies.

[41] Obviously at this stage even Einstein would have no idea that instants existed and what they meant, but would be familiar with the intervals between points or contacts.

life. It is less important than life, I agree, but then we can discuss time logically but can never know even how to begin to explain life. Another problem is that, although the intellectual and literary aspects of time have become somewhat straightforward in logical thought s since Einstein, the legends have not diminished one iota. They have rather increased as so many madmen try to start new religions---and all religions are based on the nature, provenance and eventual demise of time. So the convoluted legends of time have multiplied, making the Introduction and Conclusion longer in each case than usual. For what every educated person should know is this: once local time had been discovered making time human in origin (when Einstein and Bertrand Russell were alive), logicians, philosophers and scientists were bound to make man-made time into something so weird as to change our understanding of the nature of reality and life, divinity and the hereafter---with the consideration of the nature and future of the cosmos thrown in as a bonus. We still do not know the origin and purpose of life, but at least time, the kingpin of human existence, has been conquered: we now know how it begins or how we create it, how it passes or does not pass through nature, and how it will end--- not on the Day of Judgement which is a myth (astronomically there is only one day), but rather upon the demise of our parent sun.

The way I have chosen to organise the book also calls for an explanation, or apology. I can assure the reader that I wanted a short and simple book that most students will be able to afford and understand or even enjoy it---especially the mantra 'the usage of time is the passage of time'. But it was never going to be easy to explain time outside divinity and I knew it. So I thought footnotes should be used for thought expansions, clearing the way for the basic theory, stated as clearly as possible; but then I found that on almost every page there was need for more and more footnotes, some of which are very long. Still I stuck to my basic plan. I have done this to make what I am saying absolutely clear, not confused with other ideas. I know that whatever I do the religions will disagree with my theory of time, since they subsist on the Day of Judgement mythology. How educated and in some cases highly intelligent and well trained persons still believe in such myths I simple do not know; but even so, they should know what I am saying in the main text presented in simple terms without much confusion. However I am sorry if any reader finds that the footnotes constitute a distraction; their main purpose is to make the text stand out clearly rather than trying to make the book difficult to read just because it is about time, and time is a difficult subject. That was not my intention. As a thinker I of course value understanding and the free exchange of ideas more than anything else, and have even given strong defence of religion in the book, namely people have the natural right to believe what they like without persecution on the simple premise that a person's life is his own personal property to do with as he or she chooses. If any reader has doubts about some of my arguments, he or she should look it up in some of the notes---the answer may even be found on the same page if I have been clever enough![42]

As regards the style and the number of repetitions (the long Preface, Introduction and conclusions), I pray the reader will charitably allow that time is a pretty scary subject and so closely linked to life that everybody has his or her own strong views about it: the first man to climb Everest used methods now regarded as dangerous or wrong. I feel that the contentious aspects of time need to be stressed otherwise people will

---

[42] Being the culmination of more than fifty years of research, I have to confess that many of the pieces in this book (especially starting from Part Three) have been published before.

continue to rely on the soothing balm of tradition and insist that time is so mysterious it must be divine. On the other hand, if, like me, you are convinced that such attitudes are wrong then you have to stress your arguments even to the point of boredom and hope for the best in the name of God!

PART ONE

THE ABC OF TIME

This is a logical study of time after relativity: the Einstein notion of time that it is limited to a frame and therefore "There are as many times as there are inertial frames", together with 'his' QED theory which resulted from his paper on light-quantum and the propagation of light. But any logical study of time has to begin with the ABC of time, being the consideration of Existence, Reality, The Data of Sight, Events, 'Metaphysically fixed' repetitive motions, Arithmetic and the ability to count... (I have decided to list the arguments and statements in this book one by one to avoid misunderstanding. I do not want to be accused of the common disease of sloppy thinkers who conceal phoney ideas by means of literary camouflage.) Admittedly several of my statements and declarations are longer than others, but I try to make my logic clear even if unconventional so that the soul of Sir Karl Popper (who said he was ashamed to be called 'philosopher' because of the invasion of bogus thinkers) can rest in peace. It's all part of the difficulty of writing about the most basic, metaphysical aspect of human life besides breathing. Yet even breathing is controlled by time. I trust this section of the book will give the reader the impression that there is no such thing as the passage of time; it does not pass by. We use it to live. So there is only the usage of time.[43] Most of the things we call 'time' is just existence, motion or

---

[43] But this is where the serious arguments begin because time seems to be moving, marching all through the cosmos and causing all history on earth. Yet the relevant term may be 'existence' not time. For, in fact, only the bodies we use to reckon time are moving; time is what we get when we count the cyclical motions of these bodies to know the number of cycles any event has lasted or will last, and so forth. They are what we mechanise in the clock. Otherwise the passage of time cannot be accounted for, and its nature too will continue to baffle us, leaving the way open to knaves and human wolves to frighten the wits out of us with false prophecies of doom. All religion and mythologies are compounded

chemistry. Also time is not just there without knowing what it is while claiming that it is running all through nature; rather it is calculated mathematically from cyclical motions created from the orbit of the sun and which we count as units of time for earth use---exactly like counting cycles and saying an event lasted for so many cycles. That is why a specific amount of seconds, as our SI of time, amounts to one complete orbit of the sun, and we start another yearly cycle. Now let us look at the specifics of time.

1. Existence is the same thing as 'Being', or being in existence as a conscious human being. This is and should be the beginning of everything human---that we exist. Everybody should show gratitude to his mom and dad (they are the life in him or her) for

---

of the mysteries of dreams, breathing and consciousness, intelligence and the imagination. Underlying all this is time, the kingpin of existence, which is the reason it is so mysterious. The rest of human life is well accounted for in physiology. By the way, the clock does not show us the passage of time; the clock shows only the time. This time is discrete; so the clock ticks the units of time one by one. Time is complicated and requires serious logical thinking. What the charlatans give us is far from the truth. We get nearer to its nature when we consider it from the aspect of its passage. This is what I have done, rightly or wrongly. I hope readers will think about time from its passage and thereby realise that, as a discrete entity, it cannot pass by physically---it is used, used to live. Once this is known, time becomes amenable to the scientific method, whence it may be discovered that the actual cause of the kingpin of existence is often mere chemistry and other factors already known to science, and that real time is the most ephemeral of all things. A moment, and is gone; for longer time we have to have multiples of moments---being minutes, hours and so forth. It means there is no permanence in nature. Everything is subject to processing and will meet its demise sooner or later. And so it may not be time that causes entropy, but that entropy will come sooner or later at 'a certain time', since there is always time. This is the philosophy (The ontological implications of Quantum mechanics) that nobody wants to help me to develop. The conundrum is that since our time relies on the use of human points, the time is created by man, and what does that tell us about ultimate reality since the reality is controlled by time, or our time?

combining to bring us into the world mostly through personal pain and material sacrifices. The universe did not produce us; without mom and dad it did not exist because we did not exist. This should be the basis of all education---in science and philosophy, logic and religion (especially religion), or society and reality as I have combined them below. Thus we have existence or Being. We then have to do something to existence to get time sequences. Hence I define time as Being plus activity; it is not Being alone because it requires points and Being supplies the points: one second is a flash of reality; therefore time is a flash of reality. The other units of time are longer moments or flashes of reality. So units of time are mathematically created flashes or moments of reality in varying amounts of duration; therefore duration is the basis of time, and it is caused by a variety of events: rain, ice, volumes of water, collisions, chemistry, gravity, motion, inertia, force, etc, leading to lapses of time, or shorter and longer moments of existence. Though the major events that provide long durations for civilisations to rise and fall may be caused in the cosmos but apparently not to anybody's design because the universe itself has no time and things happen there only through random events, nobody can have any evidence to the contrary. I agree with Bertrand Russell that the all-embracing or structured time is a construction, in other words, a human creation.

2.   Reality is everything that a sentient Being is conscious of as occurring in the world around him or her, including events in the womb; once born, it is you and the rest of the world and even the universe as 'reality' versus you, or the rest of the world and you that combine to create everything else, good and bad. We can't blame the external world for everything; we always play a role and that is the burden of life.

3.   Imagery is all perceptions including the data of sight, and what happens in the mind or any sensations obtained from contact from outside the individual self. This is studied in philosophy

61

under the individuation of people. We are separate and individual selves; everything else is external to oneself. The only reliable way of sensing the external world is through the sense of sight; thus Professor Whitehead was right to suppose (originally with Bertrand Russell), that, under QED, the world of sense is not an inference as the ancient thinkers assumed, but a construction with the smallest piece of matter known as light or the quantum. Scholarship may not physically build the world, but it inspires those who do; and in scholarship we prefer sensible suggestions to imaginary 'Revelations'.

4. Events refer to everything that people do, even including their dreams, thoughts and their time---yes time. Activities produce events and therefore the sense of time or timing, the timing of events. That is as far as we can go in any logical search for the basis of the sense of time and the timing of events: something occurs, and we look forward to the time, period or sensation when it will end. From that we acquire the faculty of sensing duration; if the events recur again and again, we develop the sense of timing periodic events by apportioning duration to each one. From this it is a short step to inventing the minutes and hours of our time system by mathematics. The sense of time may not be inborn originally, but the foetus does move in the womb and knowledge does pass from pregnant mothers to their foetus genetically. This is speculation of course, but that is how all knowledge began.

5. Existence, reality and the data of sight are inseparably connected. First you have to exist; next you have to have the intelligence to want to know what other objects there are in existence besides yourself; and the only reliable way to find out is through the data of sight. These three phenomena form the grand subject of metaphysics which, in plain language, means man's intellectual efforts to know why we are here, where from, and for what purpose---also why we have to die at all, which brings in questions of love, desire, pain and bereavements.

Conditions that life inflicts on human beings in one form or another as some kind of punishment, and for which the religions and philosophers try to find answers, countermeasures or cures, so that with all we are worth we can produce the most adorable picture in the universe, which is a happy smile in a human face, especially that of a satisfied child happy about something. Against this the religious retribution moral principle is the best evidence of the creeds' uselessness. Innocent infants and toddlers (who cannot even speak) often face the brunt of natural and man-made disasters, and yet retribution is the only moral principle the religions have ever produced. Living in the world starts with the data of sight, making us aware of what is there in addition to ourselves. This includes time. Now, time! Before Albert Einstein interpreted local time as 'Time pure and simple' (as Sir Arthur Eddington has confirmed in The Mathematical Theory of Relativity---see below)[44], all mankind thought time was/is divine, absolute and fixed by God (so that a second here is a second everywhere else). It writes history and is moving towards the end of time on the Day of Judgement. Through the curse of learning we now know that it is not so; but however much we may deplore learning, it should be noted that in the absence of learning charlatans (and powerful human-monsters) impose their own ideas as divine commands to ruin human life. Very few strong men or rulers have achieved glory, wealth and other benefits for their countries without pain and suffering to some people either locally or over the border. Because of the data of sight, we also have to mention Plato. He

---

[44] The best parts of my suppositions are in the footnotes as I have explained; but the fact is time is a thorny subject and if one is to offer us the solution of the problem of time he or she should be charitably granted the method suitable for the purpose, not abused on the notorious social media. Nobody should suppose that the secret of life, if ever discovered, could be presented smoothly like the contents of the bible which have enjoyed centuries of improvements from the ablest writers and poets. The same condition applies to time.

said even what we see is pre-determined by God---that is what Idealism means. Through learning---and thanks to QED--- we now know that what we see is not pre-determined but instantly woven or portrayed by light energy; but light has a velocity. The speed of light is finite and therefore takes time to radiate from point to point as it is also matter, indeed the smallest bit of matter that can exist in our part of an infinitely vast and complex universe; and it is man-made. But if time is man-made it means we create the velocity of the material radiation by which we see the world since we can never see unless there is light. What makes Einstein infinitely great is that all this arose from his suggestion that time is secular and therefore there are as many times as there are worlds, or planets. Yet as we do not know how the world was created, this raises more metaphysical problems than it solves, so we will always need philosophers, especially to clip the wings of wayward science and technology since the politicians cannot do it---yet that means going back to some of the discarded ideas of the same Plato, proving that he really was the greatest of them all! Even the very fact that many of his ideas are still relevant is a good lesson in philosophy; it means man's problems have not changed much----life and death, illness, love and hate, wars and external relations, the economy, young and old, care of the needy, sick and jobless, God and Satan, the seven deadly sins, natural disasters, rain and sunshine, day and night, town and village life, schools, politics, politicians and corruption, the sun, moon and stars, the churches, the state and the scourge of human monsters...

6. Metaphysically regular or repetitive motions are needed for the reckoning of time; and although the nature of time remains unknown (other than my suggestions below), we now know that there is no longer a universal time. But whereas we can use anything to reckon time, like tapping the finger for personal time[45]

---

[45] But of course, in Professor Eddington's famous phrase, 'until Einstein's

, structured or constructed time is derived from the sub-divisions of the yearly cycle. So personal time cannot be known generally, but constructed time generally covers everybody on earth. The difficulty is that we tend to conflate personal with structured time, making the interpretation of time almost impossible. For example, if we're lucky enough to arrive in a black hole with our senses in good working order (which is doubtful), we could only use personal time because earth time is limited to the earth, as Einstein and Bertrand Russell have explained: earth time cannot be carried across space, particularly interstellar space. By its very nature it is basically discrete, and all those who write that it is otherwise are said to be making 'meaningless noises', and I believe Professor Eddington who said this was absolutely right. Einstein did not only say absolute time did not exist. He proved that in a fragmented world one system of time cannot be applicable to the whole world, let alone the entire universe. This is now common knowledge even among school children, although saying time is just is amounts to a belief in the pre-Einstein notion of time, because constructed time is not magically 'just is'; it is logically based on the orbit of the sun by the earth---constructed, as Russell has said, out of certain factors found naturally in the world (see below). And for those who do not know that Einstein was important as a philosopher, here is only one of the reasons---there are dozens more.

7. Arithmetic and the ability to count were indispensable for the creation of the clock. To my way of thinking, the creators of the clock were the creators of civilisation, in the sense that everything human could only have been created after the clock was invented for time keeping, and priests were involved

---

researches', the 'personal time' notion never existed. Time was 'time'---cosmic, general, fixed and divine. This notion of time would be challenged today by ordinary schoolboys as senseless.

because mankind worries just too much about death and reincarnation, the origins of life, astronomy, cosmology and cosmogony, time (which at first revolved around the Day&Night rotations) and God, because the priest were in-charge, controlling all human activities. If I ruled the world, more resources would be devoted to improving life on earth and leave the cosmos to its devices about which puny man can do very little, except perhaps the odd chance of diverting asteroids from collision with the earth, something that will not happen every day whether we pray or call God's bluff.

8. All units of time are either multiples or fractions of the second---being the basic, most convenient moment (or unit) of time---and would not exist without it; and since the second is a fraction of the year, time is based on the earth year, and 'cannot be applied without ambiguity to any part of the universe'---Russell.

9. Every unit of time is a small (culturally manageable) part of reality created with mathematics. Thus giving time amounts to donating part of your life.

10. All time is necessarily discrete because it is based on the yearly cycle which is determinate and is also divided by points to obtain the SI of time. The reason, of course, is that all units of time are and have to be either multiples of the SI or fractions of it. Those scientists who still believe that time just happens to be there (as 'just is') are in the wrong profession.

11. And since discrete time cannot march, history is now seen as the march of events not time, and so time travel cannot be possible. The idea that history is time marching on has been used to distort discussions of time for ages---"as time goes by", yet it doesn't, not physically. What is the medium of transmission? All motion is physical not temporal; we use the motion to reckon (construct) time sequences. To say reality itself (existence) is what moves time is religious and wrong, that is, towards a predetermined end. Existence does not move except individually through force or chemistry.[46] The houses, trees and roads etc,

do not move except in shadows as in sundials, which may be used to reckon time. In the absence of universal time, we can trace the origin of time as 'relation between points' without mysteries; but it makes time discrete therefore most of the ancient legends of time are no longer tenable. The reason our noble Professor Eddington almost used an un-gentlemanly language about this matter is that even some scholars don't seem to understand the new concept of time. To be fair, we have to give them time. The old idea has been around for thousands of years, so we must allow people a few thousand years to learn the new and strange notion of post relativity time in which god, religion, the cosmos and mythology play no part at all.

12. The truth of the matter is, things do not happen "as time goes by" but "as the units of time succeed one another in any situation" to build a picture of reality or part of it; it's like cinematography but not quite because physical activities also play their roles in the creation of the sense of time. If time is not cosmic and is known to be fractions of the year, it means it is bound to exist in units; anything existing in units cannot be passing by but rather is only expended unit by unit or one by one. By the way, in answer to the sceptics (whether religious or not), if the units of time were not fractions of the year, we couldn't plan the yearly cycle by units of time mathematically.

13. It also implies that time is based on action and vision, or contact and the data of sight. Units of time are portions of reality (moments of life, moments of action or moments of sound etc);

---

[46] The individual motions of objects can never be counted in their billions even in one meter, let alone in the world as a general movement; and the movement of existence en masse carrying time with it is a fantasy---we know very well that it does not happen anywhere, and cannot happen. Time in motion has to have a medium of transmission and there is none traceable. Above all existence remains where it has always been, bar minor alterations.

they successively follow one another in mathematical schemes designed to reflect reality or part of it, so that 24 hours will amount to one full day, but one hour is only part of the day. Hence it happened that the discovery of local time convinced Einstein that time can begin from anywhere and does not (or cannot) flow through nature as previously supposed. To me it means time does not move physically; rather the units replicate to make it seem to pass by---that more of its units have passed by in succession as Leibniz proposed hundreds of years ago are what we know as the passage of time.

14. As a conceptual skill, tool or technique for living safely on the planet, time is conceived in the mind and applied to reality--- that, for instance, daylight is so many hours, therefore you should not hang out your washings to dry when those daylight hours are gone; and mathematicians have made it easy for you to take this decision by merely looking at the clock. Also, time is used at the point of contact and goes nowhere; its passage is the usage. It does not physically pass by since it is only a concept. What are moving are the cycles we use to reckon time---the year and its fractions are of course always moving; but their movements are physical not temporal.[47] We use them to reckon

---

[47] The earth's orbit of the sun (ordinarily) has no meaning for man---as a temporal entity, "the year" has no significance for us; it is just a word like 'me', 'you', and 'him', except that mathematicians can use it as a long period divisible into culturally convenient units beginning with the second. When something is described as physical but not temporal it means it just happens to be one activity, an activity without attaching considerations of time to it. 'Temporal meaning' refers to the sense of waiting---"how long it was there". The psychological act of waiting is the basis of time, and I argue that this will apply to every 'Being' in the universe. From a point to another there is either a pause, a gap, or an interval, waiting for what happens next is the time interval, or what we know as 'time'. The earth gives no such gaps or pauses; rather, and cleverly, man is mathematically using the earth's orbit (the next year as from the end of this one) as one long duration which is then divided on the theory of probability into seconds, and multiples and fractions of seconds; that is what we use or know as

time through our own mathematical creation of intervals; so our time is heavily dependent on our mathematics---let me repeat this to make it clear: we use the cyclical motions to generate intervals, which we call units of time; but the motions themselves are physical; they do not constitute the actual time units; they merely help us to create them.[48] There is no mystery; tapping the finger amounts to the same thing, although tapping the ginger can only provide 'personal time', not time to cover the whole world. The cycles we use to reckon time do move, of course, because they are cyclical. But what they indicate is **'how long only', or what they are supposed to show us is 'how long in time units or units of intervals'**: one cycle means this and two means that in duration, hours, minutes, seconds, etc---that's all. That should have nothing to do with the reality being so timed. But they move, the concept of duration they give does not move, and that is what we use as time or for timing duration, i.e. daylight is out there and our timing cycles tell us that it is always there for twelve cycles (or hours). The problem is that we think

---

time and time sequences. It's all mental but planned to accord with the physical conditions of life on the planet, and often used to distort the theory of time. The actual orbit itself is purely physical. The earth cannot decide to give us a good or bad year. If the earth stopped second by second, it would be different, but it doesn't because it has no mind. Man rather has invented the seconds to divide the orbit into time, using the yearly cycle as one long duration divided into time units plus the seasons, Day&Night, etc. The important point is that the division of the earth's orbit into specific units of duration for our convenience, has nothing to do with the physical orbit of the earth. Values and meanings unrelated to human needs are what we require to establish objective reality in philosophy, that is why we have to change 'points and instants' to 'points and intervals' and eliminate religion completely from the study of time.

[48] One year means nothing to the earth, nor ten, twenty, even fifty years. To us it may be everything, but to the earth it is just a physical journey. The meanings and values we attach to events have nothing to do with the events in nature, they're just physical. Culture generates meanings and values, not the ordinary material bodies.

time too moves because the cycle we use (which is moving and never even stop moving for one moment), **is** the very reality we are trying to use the cycle to 'time', worse of all, we are moving with it! We use the earth cycle to tell us how long the earth has been there. There is nothing wrong with that, but it misleads us into thinking that time moves with the earth. This problem will disappear if we use a different object to give us the cycles with which to determine durations on earth, and thereby demystify time. At the moment the philosophy of what we call time is logically defective. Man has only recently, since the quantum theory[49], began to try and live rationally; so the defective time we have is no shame, though laughable to irreligious persons like myself, Russell and others. It's only when we answer Russell's query, 'What is measured by the clock...' can we have logically consistent theory of time. Time is the natural way things happen without theories due to natural digits, individuation, intermittence and gaps in breathing and walking. Nothing goes on or happens in a linear manner, like the conveyor belt. But all that is 'personal time'. The whole of

---

[49] By the way, it is true we have begun to live life rationally, guarded by the quantum theory (as the link between physics and reality by the senses), but the issue has not received adequate research efforts in the scientific world. For in the hands of Einstein the quantum was linked to light. We see all matter as bundles of light energy; so it means the quantum links physics to reality or philosophy. This should be discussed, and as usual in philosophy, either a single writer can settle the debate or it will linger on providing topics for philosophical research for all time---like the body and soul debate---in fact, in my opinion, the issue makes the Platonic Theory of Ideas redundant: nothing can be seen except its light radiations. Bundles of light form objects and then radiate their forms about. We catch the radiated forms as the images of things. So the Platonic theory of visual perception is false simply because the quantum theory is backed by experimental data---proof. What you see is what is there in exactly the form you see it because **it had to be there** for its radiated lights to reach your eyes for you to see it. Plato was absolutely wrong because we have proof of that in the quantum theory.

science and philosophy, society and personal lives can be built on them without theories. Theories of time became necessary through philosophy, the need for historical studies, religious interpretations of reality and the hereafter, predictions, planning and organisation---particularly of social life. These required structured or constructed time that eventually became mechanised in the clock, once the mechanics, mathematicians and logicians were set to work by the religious leaders who controlled society in those days. That is why the underlying philosophy is laughable.

15. Man does not realise that time is reckoned (mainly) in the past; the year is gone before it becomes 'a year' in time for us; the same thing applies to every unit of time, the knowing is the being of time but it does not persist and so it's also the passage of time; it's gone once noted.[50] But of course if a minute is not gone it cannot be noted, the same applies to all units of time; therefore, for cultural purposes, all time is reckoned ahead by means of the theory of probability but so fast that we don't notice it. Over the centuries this has become 'just time' because it is always there and the same---the yearly cycle will always spin the seconds, minutes, hours and so forth.

16. The important point is that time is known and used only in units; the very mathematical method by which we created the SI

---

[50] This is because it is discrete, in separate units. It would be different if it were one continuous spread. That, precisely, is the point. At the moment thinkers believe that it spreads to cover the universe; they speak of time wherever they look. But, in fact, time is discrete because it is derived with points as fractions of the year and the year is determinate. We do not know of any time system in any part of the universe except here on earth; that the same factors for creating time here are available elsewhere does not mean there is somebody there to place the points for repetitive cycles to create time sequences. This is the whole point for saying 'there is no longer a universal time', as stated by Russell, and also for saying 'there are as many times as there are inertial bodies', as Einstein averred.

71

guaranteed its existence as a unit by which all other units are multiples or fractions. The problem is that ordinarily (except in analysis) the units follow each other so closely that the impression is that time is some kind of unbroken entity that covers the whole world, even the universe. We have got to realise that time consists of individual units to understand its passage as no different from how it is used---it is spent on contact.

17. Because people do not understand the above explanation they say time 'is just is', without being able to define what it is that is just there, especially since time is so closely associated with life. And that is logically unacceptable in rational thought because time requires points, otherwise we couldn't have it in units--- and wherever points are mentioned, the hand of man is the chief suspect.

18. Thus it happened that, by using logic to analyse time, Bertrand Russell, the man with *phenomenal intelligence, found what other thinkers claim is just is to be relation or intervals between points. Sentience is required to place the points to generate the* intervals between them that we know as 'time units', like the yearly cycle. Yours truly is now (merely) trying to demystify the passage of these units of time as disappearing upon usage/notice without leaving any physically traceable trials. The reason is that the mystery manufacturing plants in the universities have now singled the passage of time as the greatest conundrum that proves there is God, which is intolerable.

19. In logic, it is found that the knowledge/usage of time is how it passes through nature. Continuous time arises because the units of time succeed one another with only a thin space between them, but to the eye they are virtually in a chain. Hence the illusion of continuous time. Last year and this year arrived (apparently) simultaneously, but they are really separate units of time. What is the quantity of space between them? I personally

do not know. But the quantum mathematicians will know it; and I wonder why they've not used this professional knowledge to explain how discrete units of time become continuous time that misleads us into thinking that time is 'cosmic' because it seems to be in a continuous stream permeating the whole universe.[51] I regard this failing as disingenuous and not fair; otherwise it is an abnegation of duty.

20. On the other hand---no philosopher will overlook this---it could well be that there is no physically recognisable gap between one unit of time and the next following it because time is a concept with only mathematically imaginary points between units. Maybe that is the reason time appears to be permeating the universe as a continuous entity, such that its passage through nature mystifies everybody. Yet there is the yearly cycle and other units of time are clearly fractions of the year and therefore also determinate, making time necessarily discrete.[52]

---

[51] And yet, surprisingly, (I must stress), many people, including some scientists and academics, argue that this quirky aspect of time is proof of the existence of divine time from which all things flow, since all things have to be (exist) 'in time'. Actually, in the absence of divine, cosmic or universal time, 'existing in time', obviously, came to be the case after time was invented or known, especially by means of the clock. Until then man just existed by means of personal time, as still happens among the beasts of the forests, something taken as the natural way of doing things. Even during the Stone Age period the clock did not exist, not as we know it anyway. The all-embracing time we could never have had without mathematics is part of the creation of civilisation. An example of personal time is this: man has two hands; to move to each other takes time, but that is not the time we can mechanise in the clock to apply to everybody and all events in society---e.g. for work, play, rest, sleep, travel, etc., for which only structured or constructed time is suitable.

[52] Time is known and used only in units. As such, it is eminently easy to analyse in logic---by simply deducing how a unit of time is created. Silent existence is said to be time going without mathematical quantification, but that is wrong. It is rather existence by means of chemistry without activity---but as soon as activity comes in periodicities arise, intervals are created, and they are accountable only

21. What is the solution? Logically if time is based on the yearly cycle then it is discrete in the sense that reality has been reduced to smaller, culturally manageable units with mathematics (known as time units 'to accord with human nature.')[53]Time in units is the same thing as reality out there--- **the units refer to the reality as it has been divided into shorter periods for cultural purposes, making time discrete, and all time is known and used only in units.**

22. Again, time in units means it is discrete, not uniform or 'even flowing' as Professor Eddington called it. It is the metaphysical source of time (as fractions of the year), that makes it strictly discrete in logic, not how it appears to anybody.[54] Discrete time can only pass by in units, like the year itself. Units of time are gone when noticed, especially as noticing it is using it---i.e. to live: to know a second means you have lived 'that' second, that

---

by time in units. Even in silent existence, to know how long it lasts requires time in units.

[53] A second is reality in a flash, so is any moment and any contact outside mechanised time; but a minute is the same scenario lasting a little longer and so forth; thus periods of activity are created to suit human nature in time, which we call 'the lengths of time in units' or the duration of each unit of time. Brief, long and very long intervals are time units we use in culture for different purposes--- all of which are fractions and multiples of the second, or any other unit regarded as the SI of time. Like language and words, many factors contributed to the creation of time in units (or time and its units). For example, with the creation of language we think of breathing, sound, the air, the alphabets, how to use the fingers to hold instruments for writing and so forth. For time, too, we think of vision and the data of sight, individuation and groups, light and the speed of light, intervals between points and arithmetic. Everything in culture had similar calculations going back for maybe centuries behind their invention. Invariably the step by step evolution of things is lost to history, giving rise to speculation, superstitions, false theories and even fraud. For now time is the biggest of them all---it used to be life before Darwin. In future it may be AI.

[54] Silent and 'un-mechanised time' is discounted because in all cases when one desired to know 'how much time' for any purpose, discrete time alone could be used.

minute, that hour, year, ten years, etc. Therefore the passage of time is the usage/knowledge of time and vice versa, because the noting of that second is its being and is instantly gone (used to live) to be succeeded by other seconds ad infinitum as continuous time so long as the earth continues to orbit the sun. Thus the study of time is the most intractable in life---simply because it is closely associated with life and yet independent without having a physical existence while controlling all activities. Nothing is more susceptible to mysticism and superstition. We can't even define the temporal length of one year without using any of its parts, and yet all units of time (including the atomic units) are fractions of the year. So here is the most annoying brain teaser: how long is one year in temporal terms or duration?[55] In the absence of that knowledge, trying to estimate the age of the universe with our years is meaningless.

23. The year does not come from anywhere with authority as a universal unit of time to determine age and ageing; it is just a physical orbit of the sun, and we can never know how long it takes, metaphysically, without relating it to something in man's crude experience. Counting the years to determine age and ageing is a sad admission that we simply do not know much about life, and yet we apply the years to the entire contents of the universe; to me this indicates that the brain is alien to this part of the cosmos and does not understand much about our lives here. The reason we get so much problems with it. And it is about time man got used to the notion that human beings and all the grand ideas in their heads are worthless in the cosmos--- yet we can create civilisations and acquire good enough

---

[55] This question is enough to reveal all time to be secular, for the phrase 'temporal terms' can only have meaning by reference to the orbit of the sun as 'one year' from which all other units of time are derived, including the atomic units which are based on the second.

knowledge of the cosmos to be able to affect asteroids with our actions. Let us ponder that.

24. However those 'crazy' scientists researching AI, electronics and the quantum, space research and super-natural events to the ultimate should know that the brain does not really care about human welfare and could lead us astray. Just following material sequences without regard to human frailty, vulnerability and delicate hold on life, could spell disaster for mankind. For instance, life is not guaranteed because the newly born (to carry on the future) dies easily if it is neglected for just a few minutes---yet 'the mighty brain' cannot help. So let nobody believe that man is born to survive and the parents of those who survive are the real Gods of The Creation of human life. The churches have known all this from the beginning of life, but for self-preservation they invent myths to conceal the truth, always using time as cover because they said it is so mysterious that only God could have created it. In fact, it is not so. The logical technicalities are complex, but we can manage. Life certainly is mysterious, but then nobody knows what it is and why we are here, except to live it and try to make success of it, which we cannot do if we are going to spend so much time worshipping the mythical Creator of life. Still, people are entitled to worship if they want to do so, without causing any harm to others. In that respect, it is a good thing that Einstein and Bertrand Russell were not persecuted for the suggestion that the discovery of local time meant that divine or cosmic time was abolished, paving the way for my suppositions.

25. **In any case, I wish to stress, for purposes of clarity, that there is no idle time, or time wandering in the universe aimlessly, just passing through nature. Time is conceptual, a mental scheme, ploy or technique, for managing life safely on this planet (the simplest illustration is not going into the bush at midnight). Knowing time is using time for something; that is how all time is known: the knowing is the usage and the usage**

is time's being and passage at the same time because it is known in units, so, as each unit is 'experienced', it is gone; and that is how time passes by---and we all know how swiftly the seconds fly pass. Traditional time got it wrong by considering it in the form of 'a blanket entity', that is the reason nobody could account for how it passes by convincingly and had to appeal to divine authority created by themselves for themselves. Yet time is so important: all civilisations depend on time, while existence is owed to chemistry, and, therefore, given quantum action, unpredictable. But to live and prosper, we have to rely on time created by ourselves as 'relation between points', thanks to Bertrand Russell. For time is not just any action, gap or interval; everything people do can be set to time, or can be 'timed'. But scholars and philosophers are concerned about how we get the time in the first place, since divine or cosmic time is abolished due to the discovery of local time. The time we use in society or culture, science and philosophy, and for all human creations, is 'structured' or 'constructed' time for planning, running or predicting all events. Ordinary intervals between events are just the natural way things happen; two trees cannot stand on the same spot, and to move from one to the other you don't need time; it is just the natural way things happen, although it will take some time—-but constructed or structured time for telling 'how long it takes to do anything?' is another matter, as it includes guaranteed existence, (that we are there)[56], planning, prediction and historical dates. All human creations and all civilisations depend on mathematically constructed time; it also cemented all religious ideas. Yet it's not worth it. Thanks

---

[56] Actual and probable time give man the sense of continuous existence. And over the centuries, due to tradition, even the probable time has come to seem actual or guaranteed. This condition of human life is exploited by all the religions in their sermons and prognostications.

to Einstein time is no longer that mysterious. For a start, it does not pass through nature[57], and with a little mathematics we can mechanise intervals in a clock; wisely we have chosen the long interval known as 'one year' to simplify the task.[58] Otherwise the millions of intervals between millions of events do not constitute time we can mechanise and carry with us as the guide to what is happening on the planet or immediate environment. It's rather just the natural way of life or 'being in existence'---gaps between 'Beings' for the purpose of individuation, that's all. The wonder is in the mathematics or the brain we use to create the mathematics.

26. let me explain what we call time again for everybody to understand: Say there is daylight out there every day. To know

---

[57] For how can time pass through nature when it is not a physical entity? It passes through the mind, though, for that is where it comes from. Of course, personal time differs from structured/constructed time: personal time is what one does with the space in his or her perspective which can be divided into time units with points if required, otherwise we just carry on our normal activities oblivious of time and space, philosophy or physics. Constructed time is the time derived from the motions of the planet that covers everybody on earth and which is vital to philosophers, physicists, economists and governments. But both are products of the mind and do not exist physically out there—that is why they can be manipulated by governments and the religions. In 1949, during a total eclipse of the sun in the British Gold Coast Colony (now proudly known as The Republic of Ghana), in many villages people were forced to go into the churches to pray for salvation because we're told the darkness (and as I recall, it's very dark) signalled the end of the world. When normal life returned, after a few hours, the priests said the prayers did it!

[58] It must be realised that in the absence of a universal time we can use any repetitive motions to reckon time (in a simple format: two cycles means this and three cycles means that, and so forth), The basic rule is how wide an area to cover, and for how long; the earth-year (or cycle) gives all mankind the widest and longest cover of all. **The importance of earth-time is that (a) it covers everybody on earth; (b) in the clock or watches we can carry it to all corners; and (c) it is there forever so long as the earth orbits the sun regularly.**

how long it lasts for purposes of planning (or whatever) we count regular motions or cycles, and say 'for two or three cycles'. But that would be personal time; and so for everybody, we use the yearly cycle. Yet that is too long, therefore we pare a year down to the seconds and all the other units of time derived as fractions of the year---seconds, fractions of seconds and multiples of seconds. The idea is mathematical and shows how our time is heavily dependent on mathematics. This has taken hundreds of centuries to evolve. It's all the fault of religion and the nature of life. It's so precarious that religion is necessary, and the religions base their false theories on the nature of time, which they regard as the last refuge of God after Darwin, Einstein and Bertrand Russell. In fact, time is far less mysterious than the origins of speech and language. So long as there is space everywhere and there is space even between the fingers, there will be intervals that can be turned to time in units with the intellectual use of points or mathematics. We are only confused and lost when religion brings in time travel to meet your grand-parents before they're married, life after death, numerous planted stories of finding dead people living elsewhere, the return of Christ, universal time that is the same everywhere throughout the cosmos because it is fixed by God, and so on.

27. I come now to the subject 'Being and Time'. Being on its own is not time. It may be argued that Being is time (a) because we have to be, and (b) Being there too we have to use time to do anything at all. It is true that Being, Time and Reality are inseparable but there are two kinds of time---personal time and mechanised or the all-embracing time; and while personal time depends on Being, mechanised time does not. It is in the clock and rather depends on mathematics, mechanics and the orbit of the sun, and therefore can be transmitted over distances and cover the whole world with scientific accuracy. A lot of mechanical processes are triggered by programmed clocks to

work even when there is no one present. Personal time is just the expenditure of ordinary energy, like moving from right to left. It takes energy and time to do so, but not something we can mechanised in the clock for general application on the planet. Of course Martin Heidegger is often mentioned to me because of the title of his book; but he was influenced by German Idealism which is anathema to me and my ilk, which is the reason Sir Karl Popper said he was ashamed to be called a philosopher. Besides, Heidegger never really understood relativity, yet I can only write the philosophy of relativity.

Let me now explain that we do not actually count cycles to get our time units on this planet. This is how we reckon time in plain language: we have found a repetitive cycle that covers everybody on the planet. It goes round and round without stopping. Each cycle takes time, meaning a certain amount of duration is expended, and yet it is the duration we need for taking any action, and which we call time. In this case, luckily, the duration is a pretty long one. We give a name or a number to every amount of duration created by one full cycle, as a portion thereof---from seconds to hours. That is how we create or generate the necessary durations for the units of time we have. This means that the only way to define the length of duration is by metric distance; yet that is not good enough. For it does not give the duration in sense. We cannot visualize what one hour is or means, least of all one year; and the same applies to every unit of time.

To show only that one second is about twenty miles is not going to tell anybody what one second is in sense. **Therefore duration can never be defined in logic, except by distance. And that is why the nature of time too can never be known by mankind, again, except by distance; in this case the distance round the sun which we**

divide into units of time from the shortest to the longest, being the year, for the purposes of guiding our activities on earth. The best illustration of this point, as I keep saying, is midnight, which tells us not to go into the bush. So all time does is to indicate where the earth is at any moment and therefore what conditions to expect on the surface of the earth. There is no better definition of time anywhere, in any language, in mathematic or in logic. Thus our time system is strictly linked to the orbit of the earth round the sun. Without the sun the system will collapse, as we cannot be counting cycles every second of our lives to generate the time by which to live. Nobody knows what creates the original duration of which going round the sun is a version because it goes to the origin of existence overall and nobody can ever tell us what it is that created the universe of Beings. What created the sun and put the earth in its orbit? Was it by design or accident even if gravity was responsible? I suspect it was a gravitational accident and, if so, then it is futile searching for the theory of everything which may even be dangerous. Gravity is caused by matter, the effect of a huge mass of matter. Light energy is also matter, the smallest bit of matter; it could therefore be possible that finding a way to unite them might lead to the absorption (guzzling) of these tiny bits of matter in great quantities by their oversize neighbour/s to the extent of endangering our existence---something like a colony of planktons guzzled by a blue whale as a snack. We don't know; but not everything needs to be known.[59] We want to know enough to keep

---

[59] We are playing with fire, especially as the military has shown interest in gene editing researches. We could end life on earth through mistakes or deliberate acts of mass destruction, even the Higgs boson is a suspect, and I am not persuaded by Cern's defence about spinoffs for seeking to spend so many billions in a world of such dire circumstances. Cern claims credit for the internet, yet it is not

our vulnerable bodies alive, even then only to the end of a very short lifespan.

---

clear that the internet is wholly beneficial to mankind. Also in electronics and the internet there are the potentials for massive catastrophic events. If we are not careful science and even medicine will cease to be beneficent but rather destroy the world through accidents, greed or mistakes, particularly as human beings are all too prone to make mistakes. Something the philosophers have missed is the increasing number of groups who are willing to destroy the world.

PART TWO

THE ESSENCE OF TIME AS CREATED BY MAN

In this section it is argued that the essence of time is this: time is the same reality as in the data of sight but reduced to long and short mathematical units linked to metric space as purely a mental tool to guide human activities[60]. As such time does not exist out there physically; we created it as a mental mechanism, and it also cannot pass through nature because it is used on contact and instantly gone, or the philosophers would say we use it to grow, as all growth takes place 'over time'. We grow with it. This gives time some aura of mystery but not as much as language and speech or the very fact of existence. However, As the earth never stops moving in the same format, the lengths of duration mathematically given to the units of time are fixed and, from the human point of view, immutable.[61] So the last minute is nowhere. It

---

[60] Man cannot sense that he is moving with the earth. The mathematicians who knew that have used the repetitive orbits of the sun to create one long duration divided by kilometres equal to time units whereby a second is equal to several kilometres traversed round the sun by the earth.

[61] The crafty religious mathematicians were quick to announce that time is fixed and immutable. In a way it is but not by God. With a combination of the earth's motions and how the SI is calculated, time is really fixed by mathematics not by God, because a second here cannot be a second everywhere else and, alas, we can create our own local time. They were very successful, though, because the system has lasted for several centuries, and even now contrary ideas are not even read before rejection so long as the universities continue to teach Moral and Pastoral Theology, something the scientists would call nonsense. God knows they have been teaching this subject for centuries; yet the world has never seen the peace and prosperity promised. When will they allow us the publicity to try something else as the scientists and rational philosophers have been suggesting? The answer, I fear, is that even for merely making the suggestion they'd send one of their well-trained villains to come and silence me for good so that they'd go on training people in Moral and Pastoral Theology for the world to remain as it

was used to live, otherwise we wouldn't be here. The importance of time in science is that time controls all activities in human life thus: without reality out there no action of any kind can be undertaken; yet with the whole of reality nobody can do anything, as man cannot manoeuvre the whole of reality in any action. However, by mathematics man is able to use manageable portions of reality in the form of time units—one minute of it, an hour, a day, et al, worked out in the mind.

Let me hurry to add that if the day & night system (which makes us human in all aspects) did not exist, we would not have much problem with time or the passage of time. Yet the day & night syndrome is merely a planetary quirk of no significance in the universe. The billions of eye-level objects and events are of no consequence at all in the cosmos; it may very well be that only the earth's gravity is felt by other bodies small enough to feel it; the mighty ones would regard it like a flick of a fly's broken limb. One analogy I am fond of is the behaviour of any ant to a human being, totally irrelevant. You don't even notice it. But such, exactly, are all planetary quirks in relation to astronomy---they do not count in astronomy even as much as an ant's scratch of the sole of your boots. Thus the day & night system which is everything to man is rather nothing in the universe. The sun is on all the time so there is only one day. The only problems with time are the intervals between events in human eyes, but they are all traceable--- caused by chemistry, force, storms, motion, accidents, delayed reactions and so forth, mainly due to our intellectual use of points. The logical basis of secular time is that (and I repeat), without points there can be no intervals, and without intervals there is no time. For consecutive time sequences, there has to be some kind of planning for the use of points that (so far) only man is capable of on this planet. But due to our ignorance, we have woven the

has always been to everybody's displeasure.

'worthless' planetary incidents into the clearly mysterious, astronomical events, otherwise even the orbits of the sun might not be noticed or play such important roles in our lives. For time involving the use of points is entirely human in origin, and the notion of destiny beyond the grave is fantasy.

So then, time is not part of the set-up in the universe. It is a human contribution (what Russell called 'a construction') using the elements naturally found in nature, especially the duration of certain actions or the consequences thereof. If this definition of time is correct, then the quantum cannot obey the strictures of time (our time, unknown in the world of the quantum), and that it is the reason we're having problems with it in certain areas of theoretical physics, including the theory of everything. Hence, ironically, the ultimate result of the logical application to the study of time is an anomaly (something we are seeing increasingly in nature. If this goes on the secrets of life will be revealed before Christmas.) Namely, the physical passage of 'earth-time' obviously does not occur; the time is used to live and grow; it does not pass through us to anywhere. Yet the long duration of the earth's orbit upon which our time reckoning is based, is certainly passing by repeatedly with the earth's orbits of the sun. Is this a paradox?

The relative motions with the clouds and stars prove nothing, as they cannot show who is moving and how; they are well-known erratic movements (up and down, backwards and forwards, and generally contradictory), whereas at all times the earth appears as our huge, solid, stable and immovable home. We rather move on it, and so it was easy to convert its imperceptible metric motions to units of duration called time for short (such as saying: 'from 1-20 kilometres we'll call it one second on earth to build earth's time system'), and so forth, fooling everybody that it came from above for religious reasons, when, in fact, they knew it was rather the product of the human mind.[62] Rulers then used this secret knowledge to control the world in various ways until the rise of science, Charles Darwin, and the rational philosophers.

1. WHAT A MOMENT MEANS[63]

Let me re-state clearly the purpose of this book. It is written to support the new theory of time we owe to Albert Einstein, Bertrand Russell and Professor A.N Whitehead among others, and in sum my thesis is this: time is reduced to metric space round the sun, and cosmic or divine time is abolished in logical thought due to the discovery of local time. All this is true and can be verified---a unit of time has become a portion of reality reduced to a mathematical unit convenient for cultural use but in the mind only. Physically it is based on the orbit round the sun. So it is conceived in the mind to suit the space round the sun---a fantastic human ingenuity, greater than anything ever invented by man; for otherwise time does not exist in nature, only the elements for creating

---

[62] Once the time system was established, rulers used it to rule the world (through religion, deception and ignorance) for all time---as they continue to do. All the religions revolve around time yet logically they don't know what it is. If you tell them the truth about time and it contradicts their religion they want to see you buried. Also the reader will find that everywhere I am moaning that my manuscripts are never even read before rejection; that is largely because of religion. Why the teachers of Moral and Pastoral Theology have to share one world with rational thinkers and scientists beats my understanding, but I am not surprised that the followers of a certain sect are always plotting to eliminate all other unbelievers. Yet for the remainder of time for life on earth, man could enjoy a good life with the knowledge we already possess not the myths we have accumulated largely through sectarian strives and phoney ideologies. It is often said rational thinkers lack moral refinement. This is not correct. A rational thinker will not punish a hungry thief more than a violent rapist.

[63] This is to assume (from the scientific point of view) that the brain is/was originally blank---like a completely new hard disk in a computer before it is crammed with data. In this case, we now know that things that happen in the womb can also be considered as data.

time do exist, but it takes human intelligence to put them together.[64] We see this clearly in how we teach infants to tell the time, when we introduce it as if it was created by divine authority. And children accept it and are able to live safely with it on the planet.[65] Now time is not seen as being physically out there, and the arrow of time notion is untenable. I repeat, time is purely conceptual but has to have a physical base for cultural purposes, and that is the orbit of the sun---particularly for living safely on the planet as it is under the complete control of the sun. Otherwise, out there reality remains the same; it does not move by the arrow of time. But a second is a flash of it (a flash of reality), a minute is much longer and so forth, to all the various units of time thus: the clock is created in such a manner that for a certain amount of distance (twenty miles or so) it ticks one unit= one second or moment, or a certain amount of duration---and duration is what the units of time measure, so that a minute is longer than a second and so on. Therefore about 31,536,000 strikes of the clock are equal to one complete orbit of the sun by the earth, meaning that for a certain amount of miles the clock ticks one second. So within that figure of 31,536,000 seconds we have the minutes, hours, weeks and months as culturally determined. That is secular time. All units of time are either fractions or multiples of

---

[64] When Einstein said there are as many times as there are inertial frames, he meant that those frames that had not invented their own times would have no time---in other words, time does not exist in nature at all, only the elements for creating time do exist. Putting them together is another matter.

[65] Human Beings grow up with this instinct to believe that time is just there. Since the discovery of local time and its interpretation by Albert Einstein, however, logicians have begun thinking that time might not be all that mysterious. They point out that time or no time reality (in whatever form) is out there; time is like counting cycles to know for how many turns it is or was there, and that the earth-year is one huge cycle pared down to the seconds from which we get all units of time, as I have been trying to explain in this book.

one second. As Russell has averred, this is 'a construction' of time by man. The earth has only one regular day, only one regular year and only one regular second out of which we have constructed our time system. And Russell was right to deduce that this time cannot be applicable to any other part of the universe. The universe itself has not yet shown that it has anybody there to create a similar sequence of time in units. Ergo, the universe has no time of its own and everything there happens through random actions, chemistry, gravity etc.

The rest of this book is just an elaboration of some of the facts and suppositions mentioned already, the most important are these: a moment (which we know as time and include in time reckoning), is any sort of human contact with nature, meaning all acts of perceiving (visual, tactile, etc.), so long as it is determinate and has to be repeated to continue---no matter how long it lasts. Now that is the interesting point, because it can be subdivided and still be a moment or part of a moment. It suggests that time is inferred from the sensation of contact; without contact there would be neither the need nor sense of time, because there can be no action of any kind without contact, and human life would not be possible. We would not need time for doing anything since we would not be doing anything. Again, a moment is recorded in the mind as a period of 'Being' or existence and is based on contact or activity. As time in social life, however, it requires theory. But without action or contact a moment cannot be defined. But these moments, which we call time units in succession, are unique---time is unique because it is strictly tied to existence, living, breathing and whatever it is that makes life different from the inanimate. To say time is gone or has passed means it's gone into the process of living life; therefore the knowing and usage of time is how it passes by. As already mentioned,

Samuel K. K. Blankson

from this point of view the Leibniz notion of 'succession' was very important.

Now, once there is life and contacts and activities connected with the life, time reckoning of some kind will follow through mathematics, physics or the imagination. Once we forget entirely that 'Being' is time and therefore time is universal (since experiments have revealed that it is not), we'll come to realise that the same process and ideas about contacts determine and develop the sense of duration between different units of time, i.e. as to which contact was (or is) longest, shortest, etc. The names and numbers we assign to them are obviously artificial, a matter of convenience and culture; this is the reason we believe that time developed late in our civilisation and development. For as to which contact is called one minute and which is two minutes and so forth depend on both the length of contact, language and our theory of numbers, all of which we invented several centuries after we became 'Complete Human Beings'. Let us not forget that we developed step by step: first we learnt to walk, talk and write. Later on we created the various implements all the way to the creation of steel and other metals, machinery and manufacture and felt the need for clothing to cover up! Thanks to the scientific outlook we have been inventing things ever since up to the computer and the electronic age, including AI, going to the moon and so forth. Time is one of them---we didn't have time or the clock when we're living on trees as monkeys. But then we grew wiser; we became human beings, engaging in numerous activities---'contact' is the kingpin as stated above. And to know the duration of activities (in everything we do) we invented a time system, being one of counting intervals between points because it is based on repetitive or cyclical motions, and we call the units 'time' or time units and apply them to determine the duration of everything we see, deal with, encounter,

89

touch, or, in a word, everything we do. It is not right to call this the most mysterious thing ever invented. In my opinion, language and speech are rather more enigmatic.

2. THE ROLE OF MATHEMATICS[66]

Our time system relies heavily on mathematics, since the passage of time is mathematical rather than physical. And this is how I like to explain the idea. First, I ask how does the year pass by to become the centuries? The answer, of course, is that it repeats the orbit of the sun; it goes round the sun again and again, because it is a unit; the year is a complete, determinate unit of time, of which all the other units and sub-units of time are fractions. As a unit, the year does not move on to two, three, four, and so on. Rather it replicates to grow to become the centuries. We go round and round the sun perpetually and count the orbits as we go along. Since all time units are fractions of the year, they also grow *(or pass by through growth)* in the same way. The implication is that there is only one year and also only one second, or whatever is regarded as the last unit of time through fractions---thus time passes by through mathematics rather than physically: the years increase in numbers to pass by; they do not pass by physically---one second is just one moment of time. Multiply that and you get more moments of time, meaning time is passing, or that it constitutes the passage of time. You can then extend the process to eternity through mathematics, physics and reality as the passage of time.

In reality, there is no such thing as the physical passage of time; there is only the usage of time because time is known only in usage. As you

---

[66] As a professional class, mathematicians played the most important role in man's creation of time sequences.

use a second it is gone: the knowing is the usage, and the usage is the passage, very intriguing in linguistics but in logic very simple. It's quite impossible to know time in any other way because it is what you use to grow, and you are always growing so long as you live.

The year is gone as it is noted (used to live), until then it is not 'a complete year'. All other units of time, however small, and being fractions of the year, pass by in the same way. So there is no such thing as the physical passage of time; there is only the living of time, or the usage of time, since we age by using time to grow as mentioned above.[67] Every unit of time is lived, and that is how it passes through nature. If you haven't lived time, it hasn't come yet[68]; but once you have lived it, it is gone, passed away. That is my answer to the problem of the passage of time. The theory of probability plays a major role in the cultural reckoning and use of time. Many other conditions and practices in society then contribute to the creation of the mysteries of time; otherwise in logic it is not so strange as to be completely incomprehensible as the religions make it---especially after Einstein. Let me repeat: time is a device for telling that daylight, for instance, is usually out there for so many hours, the hours having been derived with mathematics from the repetitive yearly cycle.[69] But this is a theory

---

[67] Time's usage, knowledge and passage are intertwined or one and the same thing: we know time when we have lived it and thus passed and gone forever. That is how time passes by. The knowledge is the usage and the usage is the passage when it disappears---always in units. The discrete nature of time is very important for its interpretation. Once it is known (or accepted) as discrete the explanation of time and its passage through nature become easy to account for in logic.

[68] In the prisons they call it 'serving your time'. If you haven't served it, it hasn't come yet.

[69] Using the hour is exactly like using a cycle (or counting cycles to determine time) because the hour is part of the yearly cycle. Through the beneficent

of time after the supposition (from Einstein) that time is not divine; as a result logicians set to work to deduce the 'probable' nature of the time we use and therefore cannot deny that it is existing, with the proviso that it is not physical but merely mental, a skill or device for controlling nature---controlling, managing, living safely etc. We are misled by the motions of the cycles we use to reckon time, for they are in regular motion; this makes people believe that they are the motions of time---the sundial is believed to be an emphatic evidence of that. Let me explain that time does not pass by because it is the same thing as 'Being' or existence. We use repetitive motions (the years, for instance), to give us an idea of how our lives are wasting away, so that ten years means you have probably only so much left to go---we mistake the motions we use for this as the motions of time. If we use the term 'cycle' the notion of time being described might be easy to understand: time is the number of cycles of a natural body. Five means 'this' in duration (say, the time it will take you to run five yards), and ten means 'that'...[70] Something of this nature is what has been mechanised into the clock---based on the earth's motions, or orbits of the sun. Note, however, that this is highly simplified, for time reckoning requires the most complicated philosophy of mathematics, especially to get it ticking so smoothly in a clock. As far as human

---

labours of thousands of very clever mathematicians we can now use sophisticated methods to simplify time into mere hours and so forth, but still using a time unit is precisely like counting cycles (intervals between points) to reckon time.

[70] They are what we mechanise in the clock as 'units of time' all the way to the centuries. Mathematics is required for the creation of time in the clock because the best source of duration for mankind (covering us all), is the earth year, but it is so long that we have to divide it into convenient units with points, which is impossible to do without mathematics---hence the units of time with which we are all familiar.

beings are concerned, mathematicians have cleverly divided the longest duration or cycle on earth (being the earth year) down to seconds or the most convenient mathematical unit. So every second is equivalent to a specific amount of space round the sun.[71] Since the orbit of the sun is repeated regularly, this gives mankind continuous units of time to mechanise into the clock. That is the mathematical origin of secular time and the closest man can get to discovering the nature, though not the purpose, of life---simply because time is intimately (almost inseparably) linked to life.

A further complication is that, in truth, other than the supposition that logically time can only be assumed to be life plus activity, as I argue, the nature of time is unknown---except as the same as being, yet you have to do something (while existing) to create time, that is the basis of my theory that 'life plus activity' gives us the theory, nature, and even the sense of time for application to reality. Even applying time alone proves that it is something 'extra' to existence, because at least sentience is required. Under divine time everything is automatically done for us, but under science no such bounty can be expected and we

---

[71] This is the reason time cannot be stopped; for it is strictly linked to, and based upon, the motions of the earth and they never stop. It answers the question 'where time comes from?' which has been used to claim that it is so mysterious that it could only be divine. In fact, time is in three parts: philosophical, mathematical and the physical orbits of the sun or astronomy. Of course, that makes it very difficult because very few of us can know something about all three subjects. Even mathematicians and cosmologists miss the philosophical aspects (i.e. local time means time is not cosmic, divine or universal), but precisely as the mechanics work it out of the motions of the earth round the sun. That is why Russell asked the question, if cosmic time is abandoned, what then is measured by the clock, the answer to which query inspires some of us to write books such as this one.

have to think of everything by ourselves---sentience, arithmetic, a theory of numbers, continuity, the ability to count, et al.

This picture of time is what I describe in my books (rightly or wrongly) as the logic of time in the universe[72], supposing that any ET Beings would face similar situations, provided they knew something about the Theory of Evolution and were trying to discover how life came about without a systematic Act of Creation. It suggests that life is not just living and breathing, but consists of several elements, perhaps accidentally cobbled together, one of the most essential of which is time, the sense of time, or the need for time at all.[73] The main thing man has to accept is

---

[72] There is no universal time---Russell. The universe itself has all the elements for creating time but no intelligence of its own for putting them together. There is no intelligence in the material universe, yet it can create intelligent beings, just as there are no babies in ordinary women yet they can create babies once the necessary chemicals came together through accidents. It means creativity rules the world but is usually accidental. So the whole of reality (including dreams) is an accident and that is that. Whatever intelligence is, it surely arose through an accident. Therefore it is of course very likely that similar accidents, higher and lower pockets of 'minds', will occur somewhere else, even in several places, in such a vast universe. There is no guarantee that such 'Beings' will be friendly and communication might be impossible. Thinking about the cosmos like we think about one another may turn out to be fatal---mankind should always remember the story of Frankenstein!

[73] **To act by time you have to have a mind that can read the time from somewhere; that is why the universe does not act by time but only by chance and accidents some of which are frightening merely to think of. Things happen in the wild that seem to obey time but in reality they may be caused by invisible factors---particularly physical and organic chemistry, inertia, force, motion, accidents etc---because the discovery of local time indicates that structured time sequences appear to be a human creation and did not exist originally. And if this local time (or our own time), is based on points as 'relation between points', then sentience cannot be ruled out and religion has lost the fight over the origin of time. This is a triumph for philosophy, but then the whole of philosophy is a field for suggestions; it is the duty of**

that in the absence of a universal time, a system of time sequences outside the human mind cannot be conceivable if it cannot be proved to have been bestowed. Nobody is ruling anything out; it is a matter of logical accuracy, intellectual consistency and credibility. If Evolution is true, then this idea must be the best logical definition of the spontaneous evolution of sentient life in the universe.

Fortunately there is always permanent contact with the earth to support this idea.[74] We can regard that as man's metaphysical connection with nature and reality (namely, freely using the earth for sitting, standing, walking, etc.); that experience plus contacts even in the womb may give everybody the sense of intermittent contacts which is the basis of the sense of time as intervals between points---**or intervals between contacts usually experienced (when single) as a gap, a pause, or a period of waiting.**[75] Otherwise we should be asking about where the sense of time (as a succession of periods) comes from? Or periods of what and for what? It is the periods between contacts. The process has

---

scientists to ascertain which philosophical suggestions are true of the world, and how they can be employed for human welfare. Thus science needs philosophy and man needs science.

[74] We can even infer this from John Donne's poem: "No man is an Island...every man is a piece of the Continent, a part of the main..." Man is never free of the earth as a source of sensations, and the sense of time is a succession of sensations, contacts, activities. To live completely without these the sense of time will not arise. Of course all that is rejected by the religions, but then we know what they are. But time is the closest thing to the nature of life, so it remains one of life's eternal mysteries that Einstein got the courage and brainwave to question it, one happy result of which is the secular theory of time and how it passes by to demystify all time completely, at least for philosophers with three credentials: respect for Bertrand Russell, Einstein and physics.

[75] Naturally, the duration can be anything; so it is safe to state that the gap may be variable, extended or subject to debate, since time means different things to different people in all kinds of situations.

95

become complex, but this may have been the beginning of the sense of time as "relation between points". That, in any case, is how the great Bertrand Russell defined time, space-time and the sense of time in man.[76]

3. RELATIVITY PLUS LOCAL TIME = DISCRETE SECULAR TIME[77]

As already mentioned, this is post-relativity philosophy at the most abstract level (not for the amateur), and, unsurprisingly, it revolves about time. But it has always been known that time is basic to life, what is new is that the time is not Newtonian.[78] The religions took it the other

---

[76] See his Analysis of Matter, Ch. XXXVIII---the whole book is a gem.

[77] Secular time which arose with the discovery of local time, is strictly based on the orbit of the sun, and since the orbit is cyclical and repetitive, the fractions which we know as 'time units' are necessarily discrete. Discrete time cannot march through the cosmos as universal time, and so all references to time in cosmology are untenable. As Russell has observed, the only time we know of is earth time---not applicable to any other body in the universe. Cosmologists are not aware that most of their suppositions are flawed because of their misconceptions of time.

[78] This may be seen as 'counterblast' to Newtonian time and A Brief History of Time, too. The ideas behind both theories are philosophically naive---scientists should never dabble in philosophy unless they're professionally competent like Einstein. To Newton, post-relativity time shows that it cannot be general (the same everywhere), or absolute because we can create our own local time; and to Hawking the reason is that time has no history. History is the march of events not the march of time, since it does not run through nature, while its very essence, too, is how we use it and it disappears upon use; it does not pass through. Time, as Space-Time, is produced with points and therefore basically discrete. Discrete time units, like the year, cannot march and they increase in numbers to pass by, so there's no mystery there. The passage of discrete time is mathematical not physical and is also how we notice time at all. Time is always passing and the passage is the usage; it doesn't go anywhere and doesn't do anything, merely a conceptual adjunct or guide to action. If I have to repeat it a hundred times, history is the march of events not time. Naturally we have to record the times of events but the time does not cause them. Several agents are known to cause

way; however, thanks to Einstein, we can now think about it in logical thought, and even school boys can tell which is the proper way: so long as we all appreciate and enjoy the products of logical thought as science, it is most unwise for anybody to choose the other way in arguments opposite to science and which we call mysticism, religion or whatever from the dark side of the human mind. I mean, there are people dealing in black magic on mobile phones, that's cheating!

Russell's 'relation between points' notion shows that he really was a great philosopher and honest mathematician; for the idea proves the secular origins of time, and therefore has been deliberately ignored by mathematicians. Everywhere we hear that man's intellectual operations are based on points and instants; yet it appears that only points can have independent existence and that the instants arise from moving from point to point and felt in the mind only. They cannot be exhibited out there physically; but from the minds of mathematicians they can be mechanised in the clock as units of time, or periods of waiting from one point to another---the most convincing logical definition of time. It proves that time is created and not bestowed, and that it is a human concept. Relation between points is the same thing as points and instants; but one is created, the other is assumed to have been bestowed---see Russell's books, The Analysis of Matter and Mysticism and Logic, particularly the latter, where he states that time is 'a Construction', Ch. Viii (x).[79] The logical challenge to time as an

---

events in nature (and the periods of waiting in our minds), but time is not one of them, because, as relations or intervals between points, time is conceptual not physical. As Professor Eddington has observed, time was completely misconceived before Einstein, and has generated the greatest number of myths.

[79] All this has been there for nearly a century yet publishers rather cashed in on A Brief History of Time, which implies that time is running through the cosmos with its own history and it is like this or that in black holes or cosmology. And

absolute and general entity, begins with relativity, as Russell makes clear in this passage from his book; and it justifies the notion that, intellectually, Einstein changed the world---even before we bring in aspects of the quantum theory for which he won the Nobel Prize, QED, and the consequent electronic revolution, gravity and modern cosmology.

At the time of Bertrand Russell and Einstein, Professor A.N Whitehead too was alive, and he was some brain, a mighty brain-box in mathematics, logic and philosophy. He defined time originating on this or any planet as post relativity time, implying that divine time could not exist. In The Principle of Relativity Professor Whitehead wrote: "...a moment of time is to be identified with an instantaneous spread of the apparent world"--- in other words, a moment of perception, vision, existence or 'Being', and went on, "...A time-system is a sequence of non-interacting moments [however that moment is defined]".[80] He was not a very lucid writer, but I suppose this is what he meant: every unit of time is a moment in life. Of course some of these moments are very long, like the year, or day, but they are moments in the sense that they

---

the writer is hailed as a unique genius. In fact, he's a unique joker or magician. Professor Eddington, who screamed that time does not flow and the great Bertrand Russell, who declared that there is no longer a universal time after relativity, must be turning in their graves with fury. So many academics are claiming to be as clever as Einstein, but philosophy is not as easy as many people would make you think---people need to be reminded that Einstein was also a philosopher; and, actually, because all knowledge is gained from hit or miss ideas, science owes a great deal to philosophers' clarifications and logical acumen. I may not have the clout of nuclear scientists, but I can suggest that the reason the photon does not obey the strictures of time is that it does not recognise time, since time is a human creation, and I have been saying this for a very long time indeed.

[80] The Principle of Relativity, Cambridge, 1922.

are determinate contacts that have got to be repeated to continue. This is an attempt to find a definition for time that suited its new status as a secular entity, created or constructed by ourselves----the year for instance, pared down to the seconds or even the atomic units of time which have always to be based on the second to make sense. Otherwise there is only one day and only one year; so any time system based on them is bound to be discrete. There is only one day because the sun is constantly shining to give daylight or what we call 'the day'; the Day&Night syndrome is a planetary quirk not known in astrophysics--- just as an individual's personal habits are unknown to the rest of the earth's dwellers. There is nothing we can only do in the night and not in the day; with artificial light we can do everything day and night. Only minor chemical conditions may be affected by the shadow called night— such as photography and human vision most of which may be irrelevant to the cosmos.

4. WHY DISCRETE TIME CANNOT FLOW THROUGH NATURE

But this is where the problems begin because discrete time cannot run all through the cosmos. Discrete time cannot bend, move or march because it does not flow through the universe; discrete time will not make time travel possible since it cannot flow forwards or backwards;[81] discrete time does not create the story of history, which is rather seen as the march of events not of time for only events can move forward, so that we carry the past with us always as the present and to the future. Thus Einstein was right: past, present and future are mere linguistic illusions: you live with your past and will carry the same to your future--- examples are everywhere, your bank balance for instance! You do not have to visit the past to access your bank balance; and the balance tomorrow will inevitably be what you have today----the bank manager will not pile pounds into your account for no good business reasons.

I honestly cannot imagine how anybody of whatever status, intellectually, academically, politically or religiously, could contradict the reasons for secular time sketched above to claim (by revelation or whatever) that time is other than the post relativity concept of it. Yet what keep appearing in books and magazines are still concepts of time in the old format: that it started from Time Zero; it just is; it increases, runs faster, slower and so on---yet discrete time cannot do any of these things; and since our time is based on the year, every year, it is none

---

[81] Discrete time is spent and gone as an individual unit of time unconnected to any other---the years, for instance. Each year is determinate, unconnected to the next or the last. This is easily proved: one second before a New Year is still the old year; one second after is the New Year. All units of time obey the same rule because they are fractions of the year and can't spare or delay any units without eating into another's periods. This is a technical point in logic and mathematics but I hope it is clear enough for the reader to comprehend the argument.

100

other than discrete. There would be nothing to call time if the yearly cycle disappeared, although existence would continue 'for some time'! But no one would know how to express the time. This is another proof that time is existence (particularly including the data of sight), plus activity, arithmetic, the ability to count and a theory of numbers---and therefore a human invention as Russell proposed. Of course, he called it 'a construction', but it is the same thing, meaning man created time. If this had been said in the dark ages the thinker would have lost his head, but at the present stage of scientific knowledge, we have accepted Darwin and QED and so can believe that man created time and much else besides.

The universe itself has no time and therefore no civilisation---or planned existence---that is why it keeps behaving like a bull in a china shop. In short, the universe has no intelligence. In the absence of planned existence there can be no logic in anything that happens in the cosmos---in human eyes, just bizarre, infinite, creative or destructive---an endless mixture of matter for no purpose. **It's like a mudslide in heavy rain; all sorts of materials are jumbled up in mud and rolling downhill to unforeseeable end. But there would be tiny organisms comfortably hidden in this. I am not surprised that we can also comfortably hide or live in a similar flux in the cosmos, since by comparison we're even much, much smaller than amoebic organisms in an earthly flux. That the cosmic process is slow enough (in our eyes and time), for civilisations to rise and fall is just our luck and ill-luck.[82] However, to**

---

[82] I think I've got it. I think I've finally got it: What is stated above is logically the correct way man should consider his status in the cosmos---absolutely nothing. On cosmic scales, even a thousand years is not much, for in terms of duration (cosmically) it does not take that long to go round the sun. The dire problem is where our intelligence came from; yet even that information is not

interpret time with complex theories so as to ascribe desirable meaning to all that is pure fantasy. Being is natural; time is man-made, since it is being plus activity, man's activity, otherwise we couldn't have it in units. Time is the intellectual basis of all human creations; so without time there will, of course, be human existence for some period but no intellectual inputs to human life, otherwise time is not that mysterious; but without intellectual inputs we would not last long.

According to Bertrand Russell it is a human construction, and Einstein added that, therefore, in plain language, there are as many times as there are planets (and I remind the reader that this is post-relativity philosophy.) In fact, it appears Einstein and Russell changed philosophy but no one is bold enough to give them the credit they deserve. But whether we like their religion or not, Einstein and Russell were the world's most recent great philosophers, because this is the age of science, man's intellectual grave showing how we will all end-up.

Above all, with his love for logic it would appear that man came from somewhere other than this bizarre universe (trapped in an alien world), but we lack the brain power to discover the source, and without that, or the gift of everlasting life, there is no justifiable reason to worship anything. Wasting our resources to worship myths since we die all too easily, when nobody has yet come back to life after death is not clever.

---

that important because whatever be the truth we will die anyway. Infants, for instance, die soon after birth without good care; that means life is nowhere guaranteed. Of course, if tracing intelligence could confer life everlasting, then, like the gold-rush of old, we'll all rush to it, yet that is nowhere guaranteed either. I fear that life is worthless without any meaning whatsoever except to live it morally, truthfully and charitably to the end...That is the only sensible thing the religions tried to enforce, but alas, without life everlasting, the power of God waned.

My guess is that certain atoms strayed into this universe from somewhere to combine with matter already here to create these frail and vulnerable beings we happen to be---definitely out of place or not supposed to occur at all. As we see, human societies die and the mindless flux of nature goes on; the same thing will probably happen when we're all gone, for life does not appear to have been planned and should never have happened. The Pythagorean supposition (transmigration of souls) sounds intellectually seductive but obviously false, although the religions stop there, but in practice it would require knowledge of a planner, designer, supervisor, et al.

I am often confused when scientists refer to something called "The Dawn of Time" unbroken to this day. And they are fond of demonstrating this by arithmetic as they count the years. Shouldn't it be "The Dawn of Existence"? According to all the authorities cited about time or post relativity time in my books, there appears to be no "unbroken streams" of time running all through the cosmos from the Dawn of Time or Time Zero---isn't that what relativity time means and isn't the contrary idea exactly what Professor Eddington called "meaningless noises"?[83] Otherwise there would have to be different

---

[83] The Mathematical Theory of Relativity, Ch,1.1., Cambridge, 1930. He wrote: "Prior to Einstein's researches no doubt was entertained that there existed a true 'even-flowing time' which was unique and universal...Those who still insist on the existence of a unique 'true time' generally rely on the possibility that the resources of experiment are not yet exhausted and that some day a discriminating test may be found. But the off-chance that a future generation may discover a significance in our utterances is scarcely an excuse for making meaningless noises". Add to this, Russell's query as to what is measured by the clock, since there is no longer a universal time, and the reader will begin to appreciate why some of us have devoted our entire lives to searching for a logical explanation of the time we have, and why we are shocked that the higher institutions who think they're more powerful

streams for the various units of time as we couldn't have one stream of time running as seconds, minutes, hours, days, months and years---quite impossible to programme into a clock. Even the traditional definition of time acknowledges this. It speaks of "The passage of existence". In a multiple stream time it would be "The passages of existences"---plainly an illogical notion, as it would be a world of intermingling streams of periodicities all passing away at the same time with different velocities or momentum. Besides, existence never goes anywhere with or without time. Increases in time are mathematical not recorded in physical meters; we live in the same houses for centuries, and even in a dictatorship the numbers of the centuries can be changed at will.

The dawn of time ties in with the religious notion of the end of time, Day of Judgement, and so forth since God has nothing to do. Usually I try to calm myself with the consoling thought that necause time has spawned numerous legends over so many centuries they have literally become instincts that no logician can ever expunge. Nevertheless, let me try: the end of time will come with the demise of the sun and its planetary systems. The dawn of time is in your own hands under the concept of

---

than God (Princeton, Harvard, Oxbridge, The Royal Society, etc., and British publishers, TLS and NATURE, the honourable exemptions), refuse even to read our manuscripts---I couldn't even say this before I was 80 and therefore no longer afraid of the knives. Colour had nothing to do with it; these guys are nothing if not colour blind. But if they did read my manuscripts then they're not even half as good as they claim to be, except that they happen to be richly endowed---that's all the difference. Yet Einstein was not richly endowed, Planck did not even know he had discovered the visual basis of matter (that the metaphysical link between physics and the data of sight is the quantum, the greatest of all philosophical discoveries), Bertrand Russell was an orphan and I was born and brought up in the African rainforest without completing primary education, and walking barefoot till I was twenty-one.

Samuel K. K. Blankson

local time. Also, under the theory of evolution life began long before we learnt to reckon time, even then, as Sir Arthur Eddington has shown (see the notes below), not as a cosmic entity covering the whole universe and the same everywhere, but as your own local time in which the cosmos has no interest whatsoever, and will end with the demise of life on earth. Indeed the cosmos not only did not give us our time but will never know of it, as it is not sentient, nor could tract all the trillions of stars and planets in the universe. Once we realised that time is our own time, our own creation, the cosmos ceased to have any presumed influence over mankind. It is too vast to have planning and direction, overbearing or controlling intelligence and physical connections with all things--- quite impossible to imagine let alone exist. For instance, how can anybody in Europe know or even think about what a single ant is thinking of in an obscure province in, say, China; yet on the cosmic scale the differences are even greater, so great as to be really unimaginable at all.

The truth is that nobody knew all this about time as a secular entity before Einstein's promotion of local time, and the religions too, craftily, misinterpreted time for centuries to suit them. However we are much wiser now: by logic we can now suppose that time does not exist in the cosmos and that everything in the universe happens through accidents, chemistry, motion, force, and so forth, which are purely material events without planning or intelligent direction.[84] But down here we cannot live long without time which arose out of the desire to regulate activities (or arose as the product of our intelligence, the origin of which we do not know). Civilisation is the bonus. Most probably we are not alone in the

---

[84] Looking at human beings and what they can do with their intelligence, if there was intelligence in the cosmos nuclear wars would be a daily affair.

universe as sentient beings, for even on this planet we have rivals.[85] On the other hand, an exact replica of man with his unique mental attributes is unlikely. The logical reasons are plain to see: Some creatures resemble man but have several attributes missing or misshaped; others have unwanted extras. It is bound to be the same everywhere due to the natural order of existence---unplanned and haphazard: when there is rain here, other places must face drought. These differences filter through to the essential nature of things, monsters, mankind---even agriculture, architecture, roads, et al. Time plays a role in all this as the agent of regulation and planning.

When it comes to the saga of life, man's life is a moving story or a journey full of incidents, that is why some good writers can write moving novels about one day's events in a single life. The Historians job is to tell that story, but they make the mistake about time; almost all of them record history as if it is the march of time. In fact, all stories consist of a series of connected events in which time plays no part. Time has nothing to do with what is happening to people except to record when they occurred, by showing that whatever it is was at a point where the sun was in this or that position---which would be translated as 'time', since we use the earth's journeys round the sun to reckon time. This in no way means time is causing the events of history from the Stone Age to the electronic age and continuing.

---

[85] I think mankind is unique not by an act of creation but by the operations of time and chemistry: whatever happened to make us what we are is long since gone and lost! Man's metaphysical loneliness is real; but don't let it worry you. We have billions like ourselves living and all we have to do to be happy is to love, appreciate and help one another.

That well known role of time in human affairs is the natural and unavoidable interval/s between events (as mechanised in the clock) and will be like this (say, 6pm) here, like that there (say, 10am) and everywhere, the time will show us only when the events occurred, the clock readings, not even why or how---at a point when the sun was at this or that position. It does not mean time is passing through nature and causing the events recorded in history. Rather there is natural processing in nature and they occur at this or that moment in time. What causes these processing is not time, and 'in time' means at certain position of the earth round the sun; the mechanical time wouldn't know the whys and wherefores of historical stories even if truthfully recorded. It is a pity that out of vanity, many historians believe history is as vital as religion for human welfare, yet it is not. Even religion is also much less important than the priests tend to think. History, however well written, can help but only vaguely in politics and statesmanship, because on the personal level, whatever the lessons of history people will always be salves to their own mentality or psychological make-up.

I repeat again as this is serious and ties in with the hereafter and all the rest of it: time is not the cause of history, and the idea that history is the march of time is mistaken.[86] History is the march of events, caused by

---

[86] As a discrete mental attribute for regulating human activities, time must have played a crucial role in the creation of civilisation but, if so, then it had nothing to do with history as the story of what people were doing, being interminable strings of events as the story of their lives. Only events can tell a story not time, which consists merely of the intervals between activities (numerical units of sub-duration derived from the earth's orbit of the sun), created to alert man of what is happening around us---as the earth moves on. The Day&Night system gives the best illustration. One word 'midnight' or 'midday' is all we need to show what is happening around us---but that is all time can do; it cannot tell the story of what people are doing during any particular phase of reality. That is the job of historians.

chemistry, motion, force, etc, and just happen to occur when the time is so-and-so, for example when the earth has circled the sun ten times and it's half-way through its eleventh cycle, known as ten and half years. Modern historians should ignore the ancient writers of history and focus on the events occurring among people the consequences of which are destined to become the causes of future events on and on forever. This will show clearly that the past is here with us today to become the present and tomorrow as the future. This has always been the course of historical narratives but the influence of ancient religious writers has led modern writers astray. Other aspects of this issue involve the past, present and future debate, so let me repeat my standpoint mentioned already.

The term 'tomorrow' as referring to the future is logically meaningless, since 'the next day' does not exist, has never existed and never will come to exist in all nature till the demise of the sun and the planet. There is only one day. The poets may have their fun in literature but in philosophy matters are different. Physically tomorrow never arrives at all because it does not exist. Tomorrow is today bar a few hours sleep. The sun is shining all the time; the few hours of darkness as the night are mere shadows and of no consequences whatsoever in nature. They may have psychological implications for mankind regarding sleep and so forth, but then how significant are mankind's psychological problems in the universe? To whom are they addressed? All sorts of things cast shadows: trees, mountains, planets, moons and the like; even man casts shadows on insects and other animals---is that significant? Something we have to sort out in metaphysics from our high institutions? In fact, the night is not a phenomenon in nature; it's only a few hours of a passing shadow, and leaves everything as it is, barring minor accidents in the darkness. But there is nothing we can only do by night and not by day---other aspects of this matter are not even worthy of debate at this level. We have better things to do and so will go on with my narrative.

It is worth noting that, by its very nature, time lends itself to numerous ways of stating or referring to it in every situation. These expressions have the effect of adding to the misconceptions of what time is. However, by showing that it is human in origin, logic can be used to strip it down to its bones, whereupon it is revealed as almost mathematical in its effects if not its nature, for nobody knows what time is---perhaps it is the reflection of chemical changes in reality.[87] Also, naturally there are always intervals between events which is what Russell called 'relation/s between points'. It is also a period of waiting and the counting of cycles (both in the mind), the last of which seems mathematically suitable for calling it 'time', except that earth time is based on the remote orbit of the sun, divided into several units that cause confusion in people's mind. In any case, I am not claiming to know what time is, only how it passes by and, of course, standing on the shoulders of Bertrand Russell and Albert Einstein, point out that the mysteries of time are exaggerated by

---

[87] Mathematics is used to measure duration into units of time, or manageable portions of reality to guide human action. The best way to describe this is the act of counting cycles---ten gives enough duration for a period, say, ten minutes and so forth. Since we cannot live by counting cycles to give us time for planning, we use our mathematical ingenuity to divide the earth's orbit of the sun with its long duration and the fact that it is repetitive, which is very convenient. From that we get our SI of time as the second, therefore all units of time are multiples of seconds, while the second too can be subdivided down to the atomic units of time as the ultimate units of duration. But duration is caused (under QED everything is caused), and we suspect chemistry, delayed reaction, force, accidental collisions, rain, water, thunder etc., are responsible. This is an explanation of time in logical thought; but everybody knows that the religions will never accept it as true and may want to teach me a lesson----I fear ignorance more than guns.

the religions whose theories of how it passes by physically and causes the march of history cannot stand up in logic----unless they want to tell the rest of us that their ideas are not routed through the strict and unbreakable rules of logic.

It seems to me that life forms (or Beings) make out reality from their data of sight.[88] If this is true and general, then reality will be different in different places, and I think time (the unavoidable intervals between events)[89] would naturally play a large part in the process. As intervals between events there is bound to be time everywhere; the point is whether ET life forms would have the intelligence to use it wisely. We think we have, but there are too many errors in our history to justify that claim. It should also be noted that intervals between events (like personal time) means nothing unless they're constructed or planned, which would require sentience or intelligence. So, in all cases, constructive time is the creation of intelligent beings. This does not preclude incidents in the wild that appear to be time-controlled, only to turn out to be caused by accidents, chemistry and so forth. It is not

---

[88] This is what Professor A.N. Whitehead meant by his supposition that the world of sense is a construction rather than an inference; but he and Russell concentrated so much on mathematics that the point was never developed. Yet, carried to its logical conclusion, it means Plato's Theory of Forms cannot be true; for we know that QED makes its own images of things and transmits them to the human eyes to result in the data of sight as the basis of our knowledge of the external world. And there are two proofs in favour of the scientific theory of vision: one is the mechanics of the QED process; the other is that the QED light can be flicked off and on to transmit or cut the images off.

[89] Time is not only known as a period of waiting in the mind or physically. It is being and everything concerned with that being, except that we have ingeniously quantified it with mathematics, but it must be understood that it includes every activity, incidents, pauses in action and everything going on or not going on even reversing, just everything because everything takes time even just for being in existence..Thus I define time as being plus activity.

110

disputed that we don't know how we got here, but we came with intelligence that the inanimate world obviously does not possess. The sad thing is that we cannot plant this intelligence here permanently because we are due, eventually, to disappear together with our sun, and the big question is why therefore were we brought here at all to suffer in vain? The religious use of human suffering failed because there is no guarantee of an afterlife, and yet it's often the good who suffer most. It is a pity, but all things involve time and the probable end of time, but wrongly. We just don't know who or what to hold responsible for man's predicament. But as Professor Eddington has averred, time no longer fits the bill because it is not cosmic or divine; and once it is shown to be secular, its origins makes it discrete, thence the deductions began, although it was not easy, since all mankind, bar a few professors, still believe that time is from somewhere and is passing through nature. Yet discrete time means time is not some sort of blanket entity covering the earth or even the universe, and of which we obtain our version with mathematics. Discrete time is created with points and therefore arrives in units, the units we ourselves give to specific amounts of duration; and if it is created in units then its absorption in usage can only be in units and the problem of the passage of time or where it goes after it leaves the earth does not arise. We absorb time to grow by it, act by it, live by it---do everything by it as the logic of time in the universe.

With reference to the arguments above, I must however stress that it's not right to say time does not exist as one answer to the difficulties it presents to mankind, even though nobody can define it logically other than as 'a construction' out of the features of the earth and other astronomical bodies (mentally constructed or created from putting certain ingredients together). But of course it means time cannot exist physically outside the human mind. One serious implication is that the universe itself has no time, and what appears to be time is caused by other factors, like chemistry, accidents, random action, force etc. Sir Arthur Eddington was a very clever mathematician and scientist of genius, the founder of Astrophysics, and he says any such ideas about time after Einstein are fatuous---"meaningless noises", as he put it. That should keep the 'Doubting Thomases' quiet, at least for now. To help them along, I give below a brief sketch of the new theory---already I have shown elsewhere that it solves the passage of time in a flash. It's all straightforward: Einstein was right. If time can be produced by anybody then there is no universal time covering the entire universe; the time we have can only cover the earth, and we have to investigate how we get this time, which inquires led us to the idea that it is discrete and therefore each unit is expended upon use so there is no need to argue about how time passes by, as it goes nowhere but into usage for growth, life and action. Man can never free himself of the legends of time; they've gone so deep into all languages that it's futile trying to change them, which is alright so long as the scientists and scholars agree that it is secular and therefore the religions should be ignored.

However, now comes professor Palle Yourgrau of the United States who is making a name for himself as the champion of Time Travel, because he has written a new book called A World Without Time (Penguin, 2007)[90] , in which he claims that Time Travel is 'a scientific possibility',[91] (which is

---

[90] The whole book is a contradiction because if the world is without time then how could anybody travel 'by time', but the publishers who claim to publish only what is 'suitable for publication by Penguin' (as they have replied to my submissions a number of times), failed to notice the illogical tissues in the book.
[91] "J.W. Dunne's An Experiment with Time (1927) caused a sensation when first published. It proposed a concept of time in which time travel seemed possible. Max Planck could not fault Dunne's maths but said that his premises were incapable of proof and so were unscientific". From The Ultimate Book of Notes

evidence that the old theory of universal time running all through the cosmos is still prevalent in a great deal of the academic world). Time defined as a moment of existence (no matter how it is perceived), whose succession creates the illusion of continuous time can be incorporated into science and is backed by experimental results, added to the views of Bertrand Russell, Professor Whitehead, Einstein, Professor Eddington and Gottfried Leibniz. However, time "as just is" (not given any definition), and which began at a date chosen by the writer (like Archbishop Ussher) is the bedrock of mysticism and has no place in science---yet scientists seem unaware of this. For fifty years my manuscripts are never even read before rejection by some publishers. The reason, I suspect, is that some people want to believe that time travel and life after death may be feasible with the mystical theory of time we have at present, so that life will go round and round the cosmos through deaths and rebirths forever. I am afraid we are doomed, not safe even in the hands of our own scientists, and Pythagoras is responsible. Professor Yourgrau's book, quoted above, which argues that time travel is 'a scientific possibility' has probably sold millions! As a direct result (and out of envy!) I can no longer agree that life is brutish and sad---it is man who makes it so through greed, ignorance, pathological cruelty, religion, unreason, wickedness, fear of death, plus the seven sins. Considering the causes of all these evils, Sigmund Freud seems to me to have been the most rational thinker about social issues in all history. For when we are friendly, loving and kind, full of gaiety and generous to one another as at weddings, life is so sweet that we all want to live forever, but due to psychological defects in all of us, we cannot always be like that---for that reason, Freud[92] is supreme. "What did

& Queries, Atlantic Books In Association with The Guardian, London 2002, p251...

Freud say? That childhood experiences, including sexual ones, are important. That our instincts are in permanent conflict with social (and specifically modern) life. That this conflict must be contained but should not be suppressed. That our predicament is hard and cannot be fully cured. That life is, at every point, replete with meaning. That 'we are never so defenceless against suffering as when we love'. That work, love and taking responsibility are the most important things in life. These are great truths..."---The Sunday Times book review, 27/8/17, Culture, p31.

Nevertheless, crucially, Professor Yourgrau has provided evidence that Einstein did not, in fact, even try to understand the Minkowski theory of four-dimensional continuum, or, in plain language, the theory that space

---

[92] This is not to take sides in the old debate about the root causes of good and evil. The philosophers have been debating this matter for several centuries. We are still none the wiser because the paradoxes in human nature cannot be resolved by the same human nature. My personal view is that, being so fragile at birth, men are born peace-loving, with several instincts and social pressure to keep them so, except that faulty wiring in the mind sways them one way or another (as psychologists suggest). The sad solution is that if it gets out of hand eliminating human monsters is the lesser evil. The research supporting this view is this: I have lived in many villages and towns with no police, doctors or prisons. When there is a fight everybody joins in to separate the fighters, other troubles will be sent to the local chief to resolve. The sick and those injured would receive many suggestions from all and sundry and would eventually be cured or die, if it was inevitable. Where any 'alpha male' needed to be restrained, he would be given to 'a strong man', who would be able to keep him under control with kind words not walls and chains. Several communities, even towns, in South America and Africa still live peacefully by this method of social control. Industrialisation with its mega-city problems and interstellar technologies is another matter altogether. I have no idea about that, being an earth man who wants to remain singularly earth man forever. If God wanted me to fly he would have given me wings, as I cannot trust other people's technologies. Anybody who believes that man will ever be able to colonise another planet is either dreaming or mad.

and time constitute one entity: that Minkowski has linked (or equated) space to time.[93] He wrote, and I quote: "Every boy in the streets of Gottingen understands more about four-dimensional geometry than Einstein. Yet, in spite of that, Einstein did the work and not the mathematicians."[94] He himself quoted this gem from David Hilbert, therefore we can be certain it is true. Yet we know that Einstein used the Minkowski formula in his general relativity. The presumption then must be that he did so just to placate his mathematical critics who were calling for him to be hanged by the nearest lamp post.

5. WHY THE MINKOWSKI FORMULA IS FLAWED[95]

---

[93] One has to admit that if the Minkowski space-time formula is true then the world has changed out of recognition, and time travel would be possible. The irony is that it is not correct yet the world has changed through Einstein's own theories of time, namely that time does not run through all nature and the same everywhere, and that our own time was constructed by man with the use of space and mathematical points, not the grace of God. Mathematicians don't like to hear of this and so pretend that my work is not worth looking at. Luckily for them Einstein did not stress his theory of time. It was Bertrand Russell who cleverly called it one of his greatest achievements---to me it is **his greatest** simply because time is basic to everything. It leads to a view of reality not yet contemplated let alone accepted.

[94] Palle Yourgrau, A World Without Time, Penguin, 2007, p6. Readers of this book, which has sold enough copies to cross continents for publishers eager to cash in (thanks to human gullibility), need help to disabuse their minds about time travel. It is most unlikely that anything Einstein told Gödel can be taken in defence of time travel when he had earlier on declared time to be limited to a frame. Also, Max Planck has condemned time travel, and he was second only to Einstein in the pantheon of the greatest in science; if he denounced time travel it is unlikely Gödel could have either the knowledge or authority and courage to promote it. People do all sorts of things to sell books, thanks to human gullibility.

[95] It is right and proper to call time 'space-time' but only in the sense that time is obtained from space not that time is equated to space with mathematics. The

The Origin of Secular Time

Technically (see below at the very end of the book), it is quite impossible to equate space to time by means of mathematics unless one relies on 'i' which Minkowski did; but then philosophically any such theory will fail to carry conviction because time is not imaginary and 'i' can only be used to represent imaginary quantities---a very dangerous thing to do when dealing with difficult matters upon which the whole of life depends---see the notes below.[96] The main reason time cannot be equated with space is that space is physical but time is only mental, a psychological tool man has artificially created for regulating his life on earth, and, as Russell and Einstein have stressed, not applicable to any other world. This means that other planets may very well have the same material compositions as the earth (this is well known in astrophysics), yet couldn't have the same time sequences due to psychological differences leading to different languages, technology, life styles. Sadly, divine time is all mathematicians seem to think about, as just is.

Perhaps Minkowski needed to know the nature of the secular time that cannot be linked to space. It is this: we know what gives the long duration behind the earth's orbit of the sun which we divide to get the sub-durations contained in every time unit. Since we need time to live, in theory, it means we are living because something caused the duration by which we survive as living beings---i.e. five minutes is shorter than

---

Minkowski attempt was not successful.

[96] In the New Scientist (2nd Jan. 1993, P.28), Dr John Gribbin claimed that it is by the Minkowski formula that the special theory of relativity is well understood by scientists. This is plainly untrue. As Bertrand Russell has said, Einstein was infinitely polite to everybody, otherwise I believe he could have explained that special relativity was already published and acclaimed. It didn't need the Minkowski formula; besides even his ict equation as forming the logical foundation of his dissertation was fatally flawed as he used time in two senses, first as i, and then as ct.

one hour and so forth, come from the long duration of going round the sun; and it is caused by the rate of the earth's passage due to its own internal chemistry. Thus it would seem that all reality is similarly caused: duration gives time and time enables us to live. What gives a thing's total amount of duration is not always clear; but we know that the long duration of going round the sun which we divide and distribute to our time units comes from the earth's own momentum. So reality is caused not by a single event which can be attributed to intelligent planning, but by a series of accidental events. For instance, we can never discover the reason the earth has this momentum to make life here possible. Ergo, there is no direct, traceable cause of reality; there is only processing in all nature. I believe Professor Whitehead said something like this a long time ago. What you see out there is not the real thing in existence but the result of processing in nature, most of the materials involved lasting for mere seconds or up to billions of years. The cause of the duration which sustains us round the sun to make life possible is unknown. But it is from that long duration of going round the sun we get our time units' specific durations to be able to live. If gravity is the chief culprit (I say 'culprit' because they haven't done a good job, have they?), then trying to link light energy to gravity is foolhardy because it makes gravity the ultimate creator and the particle of light its basic messenger, part of itself but so minute that, in the absence of intelligence, asking any piece of matter big enough to have measurable gravity to deal with a particle of light is not going to succeed whatever the mathematics, and in any case, in this area, mathematics is useless. People often forget that mathematics is a human creation.

Theoretical physics is not all there just to be discovered. In a universe of this size there is no doubt that some discoveries would be harmful. In any case we're so fragile and our needs so limited. We don't need to dig to the bottom of the cosmos to live. Thinking of colonising another planet is even worse. Yet there are people who want to wipe us out; similarly there are people who think the universe of so many billions of mighty starts was just made for us to rule. To me it's all the fault of the religious people and the brain. Why is the brain so powerful? The real philosophical problem in the world is not the origin of life but where the brain came from to inhabit our bodies and where it goes to if the Pythagorean supposition is false. And Pythagoras cannot be right because he hasn't reported back yet----and it was all a long time ago!

However, additional reason why space cannot be equated to time is that time requires points; though points imply the use or need of space, but time is in units so arithmetic and the ability to count are needed; besides, somebody must be there to place the strategic points to generate the time intervals. So Russell was right to deduce that time is constructed. The obiter of a great thinker is sometimes as important as his or her main theories---the real divine gift is to have one or two of them in any century, but the 20th Century had more than ten of them, including Churchill.

As I keep repeating, even the universe itself has no time or sense of time; cosmic actions take place through random chance, and what looks like time lapses are caused by the hidden forces that create all duration in the universe: storms, water and ice, inertia, chemistry, accidents, gravity etc. Otherwise time involves planning, but to talk of planning in the cosmos is to destroy rational discourse with religion, which is unacceptable in post-relativity philosophy. I know the problem. The term 'Post-Relativity' is extremely contentious. Some scholars believe that Wittgenstein is a great thinker because there is philosophy and The Philosophy of Science, and he is supreme in philosophy if not The

Philosophy of Science. Others like me think he is a complete waste of time and resources, if not worse. In fact Russell said he finally rejected him because he was trying to overthrow physics. To me anybody who does that is mad. For there is only one discipline called philosophy. It has to be competent to deal with everything, science included---and especially relativity.[97] Any thinkers incapable of doing so are not welcome and should go to the churches and take their friends at Oxbridge etc. with them. The distortions in philosophy have gone far enough. Technical philosophy should be as logically close as possible to theoretical physics; it has to be because physics in the wrong hands can be used (easily) to destroy the world. Ethics too should be grounded in psychology, sociology and biophysics, for we are chemical structures and chemistry is based on bio-chemistry, which, in turn, relies on Bio-physics---although I agree that a little bit of religion should be incorporated in ethics for the consolation of weak souls. After that, all that mumbo jumbo began by Kant and others and the whole of linguistic philosophy should be relegated to literary studies---that is, in an advisory but not in a ruling or dominant position in philosophy--- philosophy is too important for that.

---

[97] The importance of relativity is that it is after Einstein's paper on light that we were able eventually to create QED, or the interactions between quanta and electrons which now control all physics and the visible world which is our part of the universe. This we can now pursue consistently and fruitfully; until then physics did not even know the true nature of light or how it is propagated and interacts with other forms of matter. So people display ignorance when they describe relativity as just one of the many aspects of physics. It is the new physics, and we owe it entirely to Albert Einstein. The same man took just one look at time and declared that it's not divine but created by man for the sole purpose of living safely on this planet.

119

The Origin of Secular Time

The creation of space-time, being the merging of space and time (still as separate entities), was achieved in special relativity, as Russell has observed, by the use of the 3+1 formula, but mathematicians were spitting blood because time had been made secular at the same time. They said it means man (as emotional, biased, partial and fraudulent as he can be!) creates his own time and then adds it to phenomena and call it objective reality. That's not an acceptable concept to represent true reality, they claimed. ***Yet 4-D geometry does not and cannot exist without inaccurate use of 'i' as well as the flaws in the transformation of coordinates,[98] so the theory of curved space-time is flawed, and time travel via Minkowski is not logically tenable.*** The transformation of coordinates is really at the heart of the problem[99], and in describing the Minkowski contribution Einstein used the words 'realised' or 'at least thinkable' which goes to prove that he did not take Minkowski serious, for ***thinkability*** is not part of the objective study in science---see page 56-57 of Relativity by Einstein, Routledge ed. Routledge Classics, London, 2001.

---

[98] Similar misuse of archaic formulas underlies many suppositions in physics; sweeping them away will liberate the subjects of many 'false' theories.

[99] Can it ever be mathematically guaranteed that a coordinate will be there even if it is not perceived by any means? In the case of time, it is even a conundrum because time is not universally permeating the universe---there is no universal time, but it seems Minkowski (like everybody else) was not aware of that. Also his famous ict equation, $\sqrt{-1}.ct$, is wrong because it mentions time twice, implying that there is universal time plus our version of it. Otherwise if the square root of minus one is representing time (as stated by Einstein in his book, Relativity, Page 58), then what does the ct stand for? Are there two types of time? Of course there is no time but 'The Time', the cause of which must be purely physical; for there is nothing left after physics that is not physical, though we human beings are quite incapable of tracing them all. But we know our time is caused by intervals between points---the year for instance, whereas all units of time are also fractions of the year.

Samuel K. K. Blankson

The history of the Minkowski supposition to equate space to time and vice versa is interesting,[100] even before we come to the story that Einstein did not understand it, for saying that in reference to any subject in physics or even science generally, means the great man thought whatever it is was nonsense. But Professor Eddington called 4-D Geometry arbitrary and fictitious (though useful for the study of phenomena in his Mathematical Theory of Relativity). Russell said it was compounded for the convenience of mathematicians (The Analysis of Matter). And one reference work (at least) called the theory artificial (The Routledge Concise Encyclopedia of Philosophy). It is true, of course, that Mathematicians adore the Minkowski theory because it enables them to dispense with the 3+1 formula, which they regard as less than objective for science.

Let me stress that the Minkowski formula is not some sort of a harmless, innocent mathematical game we can play for entertainment or to amuse ourselves. To equate space to time is so serious that it carries with it so many, many, and many implications in science and philosophy that the nature of reality itself is affected, altered or even distorted. For a start, it means that objects do contain time in their inherent natures simply because they occupy space; so time travel through the curvature of

---

[100] Equating space to time is recognised in logic as the greatest achievement of the human mind if it can be done; so mathematicians who are not competent in philosophy should not dare to look in at all, least of all comment on the matter. The 3+1 formula is not strictly objective, but it works for our little efforts to understand nature, which is what the Platonic simile of the cave implied--- namely, man can never observe **real** reality only the shadow of it. The main problem with human beings is that we know we are worthless in an infinitely vast, strange world and yet want to conquer it, though we are fragile and easily killed off! It's all the fault of the religions.

space, worm theories and many other fantasies become conceivable, and so on and on and on.

However, in my judgement, I suspect Einstein did not even bother to understand 4-D geometry. I can imagine what was going through his mind. The mathematicians had to be placated and come to support his theories so that he could get on with his work. For that purpose he praised Minkowski and pretended to adopt his formula. The whole world was misled into thinking that, because Einstein praised Minkowski, the latter had a hand in the theory of relativity. The truth is that he tried to contribute to it but failed because he had to base his theory on imaginary time coordinates. In fact, the Minkowski formula was irrelevant. There is no way it could have helped the new theory of gravity, and I would bet my last penny that Einstein knew that.

The definition of time was a different matter. Einstein said (and I quote from Abraham Pais' excellent Biography, 'Subtle is The Lord...',Oxford, 1982, CH. 7): "All that was needed was the insight that an auxiliary quantity introduced by H.A Lorentz, and denoted by him as local time can be defined as 'time', pure and simply". The many times this is quoted in my books reflects the stubbornness of mathematicians. It means local time has been discovered; cosmic time does not exist; divine, absolute or general time covering the whole universe and the same everywhere never was; time becomes discrete and cannot pass by physically. These new ideas about time were good enough for the great Bertrand Russell, good enough for me, Albert Einstein (the unique scientific genius), Professor Arthur Eddington (the founder of astrophysics), Professor A. N. Whitehead (who taught Russell that the world of sense is 'a construction' not an inference[101]), all these added to the rest of mankind who follow their ideas.[102] If, for some reason, it is not good enough for the mathematicians, the churches and the concept of reincarnation, that is just too bad, and I feel sorry for them.

Of course time is important, for it is nothing but life itself in action (life plus activity). There can be no reincarnation without a future and a past still existing as it was experienced. Yet under discrete time there is no future because it is constructed by time yet to come to exist through the actions of the living; similarly there is no past still existing 'as it was experienced' because it is what we have with us as 'the present'--- carried over from the past otherwise how could we be living in the same buildings? All religious ideas are easy to refute because they are based on shallow philosophies. Yet humanists have got it all wrong. Science is more dangerous than religion. What we have to emphasize is that under science we are obliged to examine ideas logically, philosophically and scientifically with human welfare in mind before implementation. For a start, this means that war must never be considered---probably impossible but it demonstrates our humanity clearly, at least in thought.

---

[101] 'A construction' implies that man constructs reality from the data of sight or through the mechanics of QED. 'An inference' means reality is hidden but the mind can infer it, or bring it out. There is a world of difference against religion between the two theories of existence, because the scientific theory based on QED goes further to the extent of being capable of destroying the world. However, all previous ideas based on inference render philosophy before Einstein and the quantum theory scientifically inadmissible, except that the classics contain words of advice and wisdom that may be helpful to man in many ways.

[102] The reader may not know all this because the subjects are so difficult that publishers prefer to go for the easy options for the sake of making cheap money from shallow books like A Brief History of Time, then call the writer a unique genius, and demand that he be given the Nobel Prize for physics. I call this the abuse of Free Speech, but others call it Democracy, the magic word in whose name more sins are committed than the bible ever prescribed.

The Origin of Secular Time

6. WHAT IS MEASURED BY THE CLOCK AS TIME?

In the absence of a universal time, Russell demanded to know what is measured by the clock. Actually, there is nothing. So what is time?[103] Nobody knows what time is, but by using the Lorentz concept that time can begin from anywhere, with Einstein's support that there are as many times as there are inertial frames, we can logically infer that time is constructed by man and that it is purely psychological and cannot exist outside the human mind. Mathematics is no help here. Generally mathematics is indispensable in physical matters not in matters of the mind.

Yet, in the end, analysed to the bottom in science, logic and philosophy, time is basically the counting of physical cycles as years pared down to seconds with points[104]; therefore it's mostly psychological. As a physical entity time is unknown. We all talk about time and use it daily, but what is it? Everybody claims it just is----but even then as what? It certainly is

---

[103] It is logically correct (even unavoidable) to equate time with 'Being' or existence; the problem is that time requires points because it is known and used only in units. So time is much more than just life or existence. We must therefore recognise that it is life plus activity: you have to be, and you have to be doing something to need or create time for your purpose. Hence the local time concept is intellectually as important as the creation of the wheel.

[104] Time is existence plus activity and therefore obviously artificial---a system invented for knowing how things, actions and people last in the world or get on. So, first came the world, then man, perception and action followed before time came to be used to account for them. Yet until the Lorentz local time concept we mistakenly thought time was eternal and divine---many people still think so, wrongly. Some people even consider me insane to be worrying myself about the philosophy of time, which they regard as 'just there'. However, Lorentz and Einstein did not find it 'just there', and studying how it got there may bring man some benefits. Is it even possible that the evolution process itself is controlled by time, **which is our own time---that is, in the absence of a universal time, and, if so, what are the implications of that for science and philosophy?**

not 'Being' because it requires points supplied by 'Being' or existence. It is also not motion because motion is multitudinous and not at all in one direction anywhere; it also has to be divided to provide the intervals we know as 'time', for which reason the motion will have to be cyclical---the year for instance. Whoever does the dividing is the creator of time---the intervals between points that give us the sense of 'a period of waiting'. Time is not that mysterious under logical analysis, thanks to Einstein and relativity. The situation is this: local time has been discovered or created by man. So it means general, absolute, divine or cosmic time is abolished. Once that has happened, we had no choice but to keep searching for convincing theories of how our 'local' time is created or began, and that is what some of us have dedicated our lives to search for. Russell's query is probably the cleverest and most important question in metaphysics ever posed. All the rest of us can do is to find the answer. But there are so many aspects of time so mysterious and susceptible to legends that no writer or thinker should incur our displeasure for advancing theories of time that seem to defy logic and the scientific principles under QED that rule at least our part of this vast universe. For we simply do not know what time is, except that, thanks to Einstein as I keep saying, and the discovery of local time, we now know that the all-embracing time originated from the human desire to regulate activities; therefore our time is limited to our planet, human in origin, and will end with the demise of the planet and human existence.

However there is another serious question. Can anybody recognise time physically? Those thinkers and writers who claim that time is passing by (and could even carry it to a black hole), should be able to identify it in physical form, yet nobody can do that.[105] We think we can but it seems

---

[105] Of course there is personal time (like tapping the figure) but it has limited

we are mistaken. We are fond of calling the physical cycles we use to reckon time (like our orbit of the sun) as 'time', but that is wrong. They are all physical. Real knowledge of time is its passage and how it passes by is how we use it, **then it is gone---gone in units (the units in succession, as I have pointed out already).[106] Since time does not stand still, the passage is the knowledge and also the usage. It took me more than fifty years to work this out.** Furthermore, time is known and used **only** in units as the fractions or units are derived from the repetitive yearly cycle. In other words, the year from which all units of time are derived is determinate; therefore time is known in units only, and since these units are not physically touchable, they must be mental.

---

applications---it is not time that covers everybody on the planet (and beyond) and by which we plan life's complex activities for living properly on the planet. Needless to say, we normally jump from personal time to the all-embracing time and confuse the two in many instances. The big question to exercise the lofty minds of future thinkers is whether personal time, being instinctively part of the mental set-up, inspired inquiries that resulted in the creation of the all-embracing time by which we created civilisation. Personally, I think it did, due to the data of sight. Human beings see things and wonder how far they are, what they are made of, where from, how big and so forth and through such inquires discovered the all-embracing time. All this is mere speculation; but that is how all reliable knowledge began. We probably started our astronomical speculations from the moon, the sun and asteroids---even the clouds may have had some influence. In any case, personal time has never raised any problems in philosophy, for it's almost completely useless. We cannot even express it without using the general terms and units of time derived from the all-embracing time---like seconds, minutes and hours.

[106] One hour is gone but another is following; a minute is gone yet another follows; a second is gone, another is following. Without the succession of the units of time there can be no continuous time, nor could the passage of time be explained in logic without appeals to the cosmic time we have recently abolished due to the discovery of local time. This is the new theory of time---logical, scientific and easy to understand.

Works of this nature are never beautifully written as much as the thinker would like, since they deal with subjects very difficult to write about at the first attempt. So, rather intransigently (somewhat), because I can't help it, let me add to my literary crimes by stressing that time does not move, rather it increases mathematically. There is no passage of time; there is only the passage of events, misconceived as 'the passage of time' all these centuries. For time consists of mathematical concepts we work in our heads and apply to nature for our benefit. If it is midnight people are not allowed to go into the bush because...The all-important ideas are implied in the term 'midnight' **deliberated in a mature mind that can understand it,** otherwise it means nothing, for instance, to a toddler. Without such mature minds there is no time and things happen only by chance and chemistry. By time I mean mostly time in the clock. Man should never forget that he grew from something inferior to what he now is, thanks to his brain. The task for the future in science and philosophy is to find the source of the brain not God---for I have no doubt that it is material, a unique combination of atoms, most definitely including oxygen, hydrogen and carbon. The issue is simple: with a material search you have something to deal with; with divinity you have only the dreams of corrupt humankind.

To get time in the clock in the absence of a universal time that can be attributed to divinity, we simply count repetitive cycles and say that 'something' (any event) has lasted so many cycles, the most convenient of which is the yearly cycle---that is time, provided we realise that every time unit is part of a repetitive cycle in nature; that's the only way to generate the 'universal' time intervals we know as time for short. But that is why time is known and used only in units. The number of the cycles (or time units/intervals) may increase; that is all. The mathematical increments of the cycles constitute the 'movement' (or

127

passage) of time showing that the event was there for so many cycles, meaning so much time which is always in units---and since these time units are known **only** by how we use them, I repeat, the knowledge/usage of time is the passage of time by means of mathematics not physically.

The mistake is assuming that the motions of the cycles that provide the time units or intervals are actually the motions of the time itself. **In fact, they are not; rather we count the cycles as the number of times an event (or non-event) has occurred.** It is the truth about this little matter that took me so many years to figure out. Because, to me, it is a little matter upon which the whole of existence depends since time controls everything. Thus the passage of time is mathematical not physical, all because time (like the years) consists of units, and the units (again like the years) increase in numbers to pass by and seem continuous. Understood as such time and the passage of time are relatively easy to explain.[107] Take an ancient monument for example. It may have been there for centuries simply by counting the cyclical units we call 'years'. The centuries have not been moving the monument; the years are counted (as centuries) only mathematically---any dictator can easily amend all ages mathematically for political purposes.[108] Time does not

---

[107] However this knowledge came to me after miles of discarded A-4 sheets some of which I have preserved, plus several dozens of rejection slips.

[108] On the other hand, if time did specifically and strictly move all things forward, then the monument would be in a position that even the little Napoleon could not falsify; yet time does no normally move things forward, some things do change (mostly through chemistry and accidents), but not everything; normally we count the units of time and apply them to objects and events, and how we count them is important for all sorts of reasons. But it is certainly true that the ravages of time have spawned numerous beliefs. They say it even causes entropy, and gravity, too, affects time. I tend to think it all depends on how time is defined or perceived, and that if there is no longer a universal time, then,

pass physically, only mathematically. The earth goes round the sun again and again; that has nothing to do with time; it is a physical journey. However, if we are counting the orbits as units of time (as years) then it is the mathematical increments we have to worry about---because as they increase time is passing by as the earth upon which the time is based is also passing by. But the mathematical increments and the physical increases are not one and the same thing because our time systems vary—i.e. one orbit is one year but we've divided it into several other subunits. Since we cannot carry the earth's orbit in our pockets or on the wrist as watches, the second in its numerous forms, as time units, constitute reality in miniature, in units we find culturally convenient.

We are using the earth's orbit as one long duration divided into the units of time we use. There is no time in nature that we can just pick up and use, but we can employ any cycle to time and regulate our activities (as 'time') otherwise we can't live.[109] That is how miserable we are. Creating time was man's crafty way of trying to survive; without time there would be no civilisation. All those grand theories and sermons, moral and pastoral theology, are just bunkum. Given that the earth's motions

---

thinking along those lines, many of the mysteries will eventually be resolved. This situation has been encountered several times in human history, most notably with Darwin's theory of evolution; the same thing will happen to Einstein's theory of time---for it was he who said specifically that there are as many times as there are inertial frames.

[109] Once a logician or philosopher is told or convinced that there is no longer a universal time, the speculation is endless; but there are strict rules to regulate philosophical arguments. This is where Aristotle went wrong. He relied so much on teleological arguments instead of simply saying 'we don't know'. When we get rid of Aristotle on this score, Bertrand Russell is right---all philosophy becomes something like science, that is, strictly based on evidence. We can use teleological arguments so long as we call them physical and organic chemistry plus accidents, if anybody can plan them.

create variable (largely dangerous) conditions through which we have to negotiate our lives, we are forced to rely on the yearly cycle for timing our activities. What we need to understand is that all units of time are fractions of the year, and the year too is cyclical and so all units of time are part of a repetitive cycle and equivalent to a specific amount of space derived from the yearly cycle. How do we convert the physical orbit of the sun to time sequences or moments? I imagine it arose like this: every nudge in the orbit is a physical unit of space (which is actually the case), and means life out there is so-and-so---all learnt from experience. The imagination of the mathematicians then took over and created the clock to make life easy for the rest of us, converting the physical journey to temporal sequences or time we can carry with us wherever we go. Thus a unit of time is a portion of reality in the mind as well as in the physical world; the two do not agree because time is conceptual; it moves only arithmetically and subject to human manipulation, besides as a concept, it cannot move over space physically. In logic time is simple: we count cycles and say, the duration of an event or something is or was so many cycles; what makes it difficult are two things. One is that the units of time obscure its cyclical origins; and two, we conflate the motions of the cycles with the time itself---the number of times the cycle (not the time) has gone round. It is the cycles we must count as the duration---the time---of the event. Of course we don't actually count cycles when we use the clock; yet time units are cyclical because they are derived from the repetitive year and wouldn't exist without the yearly cycle. Every second is part of the yearly cycle and all other units of time are either multiples or fractions of the second---no year is complete if a single second is missing. When we are counting in a New Year, we go to the last second; the next one is the beginning of the New Year.

The number of times the earth has gone round the sun are the years. These years do not move like the yearly cycle itself; they can only advance arithmetically. Almost all the confusion about time in logical thought comes from this error about the physical passage of time, as if time is running through all nature like the Gulf Stream, which is what aroused the ire of Professor Eddington so much, especially after relativity.

So, again, let me repeat, almost everything about time is mental not temporal or physical and a lot of what we think we know is misleading. The physical cycles we count as years are the only substantial aspects of time overall, since the fractions of the year (the units of time) are mental and not physical. But having known this we can use anything, any cyclical motions, to reckon time---it's a confidence trick; the real nature of time can never be known, other than the elements that cause us to endure periods of waiting in any activity. I must explain that I am not repeating myself, rather I am emphasising the points because time is notoriously enveloped in myths. So I want what I am saying to be stated as clearly as possible. God knows that whatever I do the religions will say I am mad; but if so then I want to be as clearly mad like the religious thinkers calling their dreams 'revelations'. On the other hand, I trust we can use logic to demystify time and how it passes through nature.

Ten orbits of the sun are ten years. That's true, and in the absence of a universal time nobody can define time besides the mental notation of the orbits as time (the years of course are our years, the measure of our own age); _but crucially somebody must be there to set the points for the yearly cycle or there will be no years and seconds derived as fractions of the year._ So it is also true that we create the years or time, as Russell has pointed out, as relations between points.

The Origin of Secular Time

The theory of the new concept of time is that any cyclical or regular motions divided by points will provide periodic intervals or time units for cultural use as the logic of time in the universe; that is how we get time to put in the clock. We can simplify this in mathematics thus: RM.P =TS + E (or RM.P+TSE=time), meaning any "Regular" "Motions" "Divided" with "Points" provide "Time Sequences" as the logic of time in the universe. Beyond that there is nothing for the clock to measure as time units. The 'E' represents "Existence": Regular Motions Divided with Points provide Time Sequences for defining Existence---defining, justifying, legalising etc. The logical reason is that (culturally) everything in existence has to have its "when" (or time) of existence in the universe. For while there is no universal time, there is nevertheless a universal law for time and existence as the definition of life: every 'Being' has to exist or be covered by time so that it can be defined as existing at so-and-so a time or it never existed. Every existence can only be quoted in time, otherwise how can it be cited? This being so, the above equation becomes the equation by which all life can be defined or justified either in law, philosophy or science. That's how important time is, and we can say that it's deduced from Einstein's ideas.

It must, however, be noted that time is backed by, or based on, duration, the real unknown mystery of time, which enables us to tell that one hour is longer than one minute. The logic of time in the universe arises from the fact that duration has got to be divided to obtain the culturally indispensable units of time, since all time is known and used only in units. The word time means nothing in culture without quantification---the quantity of time involved.

Duration (that is, during the period a thing is there or is encountered), of course, is natural, existing all over the universe and (under QED) it must be 'caused'; but to exist (or live in an inertial frame and have culture) you have got to find a mechanism for converting duration into time sequences, dividing it into manageable units. So duration caused by many factors (inertia, motion, atomic and nuclear processes, ebb and flow, force, obstruction etc.) must be existing all over the universe, but it takes the human mind to invent time sequences out of it. Luckily for scholars, for once it is plainly not a chicken-and-egg question since the factors, conditions and parameters for 'constructing' time sequences are all present throughout the cosmos. But since the construction requires the intellectual use of points, we had to come down from the trees to learn how to do so. The year, for instance, is determined from one point to another: from midnight 31st December to the next midnight 31st December and then start another year on and on forever. Even one second before midnight is this year, one second after and it's the New Year---thus it must be recognised that the second is part of the yearly cycle. Intelligence or sentience is required in time's construction, plus a theory of numbers, the ability to count and arithmetic. Time does not exist outside the human mind, which, of course, means another line of inquiry must now begin!

Thus time, as important as it is, is an artificial contraption based on existence for the justification and regulation of that existence and all activities. Let us say we could use the hand round and round without muscular strain. Ten cycles means it is time to go to school. At school, a hundred cycles means it is time to go and play, another hundred cycles and it is time to go back home---something like this suggestion (as a mechanism in a clock) can be seen as a device for reckoning time to regulate activities. Existence or Being seems inseparable from time in that everything we do can be ascribed to the expenditure of so much time; but that is not all that mysterious, because time is created by the critical and indispensable planetary conditions that sustain life, so that life cannot move without these conditions, and as a result cannot move without the time they create for us. Like language time is secondary to life, something that helps us to know how well and safely to live on the planet. Take the daylight for example. All we need is something like the sundial, where the positions of the shadows of astronomical bodies (or whatever) enable people to know how to go about their various activities. This should not be interpreted as the motions of time; that it moves from morning, to mid-day and to the evening and nightfall---not at all. Only the shadows do move, but it has traditionally been assumed that the motions are those of time itself. That is wrong. Time does not move but repeats its units to pass by. I have to stress this point, for there are no days in nature at all; what we experience are temporary shadows; so there cannot be days moving from morning to nightfall. From the sun's point of view, there is only one day and it is on constantly without nights.

Rather time consists of the units or portions of the earth's orbit of the sun we have pre-determined in mathematical units and apply to events. Thus all time has astronomical roots, and the motion or advancement of time is achieved through the replication of its units: one hour means the second has been repeated 3,600 times, and so forth, not that the second has physically moved through space to become one hour. Time moves through replication not physically because it is known and used only in units. Only the units multiply---the years, for instance. The result is that man is confronted by two quandaries: (a) it means we can only know how time passes by and not what it is---the usage of time is the passage of time; and (b) people who cannot comprehend this regard it as one of the riddles of time that confirms their religious beliefs about the 'mystery', and this belief is so deep that nothing can dislodge it, which is sad, because motion, events and everything else do not constitute time; rather we apply units of time to them; these units of time (or space) are obtained from a breakdown of the earth's journey round the sun, and mechanised as units of duration---seconds, minutes, hours and so forth, so that 31,536,000 seconds equal to exactly one orbit of the sun as 'a year'---every second is equal to a certain amount of space.

Let me stress that the passage of time is how it is used; alternatively, how it is used is how it passes through nature: unused time is not known; for how we know time is how we use time. Implied in this theory is that all time is known only in units; mere being is not time because it lacks quantification or the amount of time involved.[110] Being is either chemistry or physics, not time; rather time can be applied to determine its duration. Duration is the important thing and time is merely invented to apportion specific amounts of duration or intervals between points as 'units of time'. And that, of course, can be mechanised into a clock in answer to Bertrand Russell's query---'what is measured by the clock in the absence of a universal time?' For instance, if a speaker is given ten minutes, speaking for ten minutes is the passage of ten minutes, and we say 'ten minutes have passed'---the use of the time is how it passes by, and so there is no need for any abstract theory to account for the passage of time. We do not know time except how we use it, and how we use it is how it is passing by. The idea that time exists even in the wild is a myth[111]; so is the suggestion that it is the most mysterious thing

---

[110] To know that we had to invent time, namely by merely counting cyclical motions and applying units of them to events as time units; and obviously the best cyclical motions we have found are the years pared down to the seconds and fractions of seconds----that is time demystified. But it means there is only one second, and we either multiply it to infinity or divide it to the bottom.

[111] Time requires the intellectual use of points. Without points there can be no instants, moments or time---thus sentience is required. The factors, elements,

in existence. Tapping a finger to indicate the passage of intervals is also a form of time keeping. The 'greatest mystery' accolade belongs to the brain and intelligence (taken together as one entity), which gives us the power to think and invent things, but for what purpose no one can tell, since we remain vulnerable and easily killed; yet coming back to life after death remains unproven. This is the total story of life and it is not pleasant: there is no purpose for life, yet it is painful and ephemeral, giving birth alone is horrendous---so what for? Pain may not exist in the universe of massive bodies, but for man it is our biggest worry. Where does it come from? We don't know; but since intelligence also comes from the same brain, the suggestion that 'everything' (including everything, however bizarre that happens to human beings), is part of the brain's efforts to maintain order, chemical equilibrium, symmetry and stability, may not be all that fanciful.

Anyway, duration is oppressive, because the earth's orbit (upon which it is based) is strictly fixed, repetitive and unstoppable. Otherwise there would be no duration and no time. That is why the universe is not regulated by time but only chance. **Time is not the activity but the sense of duration of activity (derived from the breakdown of the year), which we apply to events, even including passive activity like sleep.[112] That is**

---

parameters etc, for creating time may be everywhere, but intelligence is required to put them together. Surely there is something which directs this intelligence and it is either God or the instinctive urge to survive. I think it is the urge because disease/injury is noted in the brain as mal-function, so there seems to be a mechanism for maintaining symmetry and equilibrium.

[112] But there is no everlasting duration. All duration is caused by temporary events. The earth's duration we use for reckoning time is caused by the temporary event of going round the sun, and in that event it is fixed by nature. Hence the time it gives cannot be changed by man, giving it the appearance and authority of cosmic control.

why the sense of time and timing grew out of activity. Without activity (general or repetitive and meaningful activities), there is no need for time and timing---the timing of how long they last. Thus a moment is 'a moment' of activity, meaning contact of any kind at all. Hence sentience is required for the construction of consecutive time sequences; yet that is created by intelligence, the source of which man can never find out other than the speculation that the brain accidentally got relevant compounds together to create intelligence, as is mirrored by the manner AI has developed, whereby algorithms can get out of hand to create problems and unintended effects.

However, tradition, religion and the sheer force of habit preclude understanding of this new theory of time that arose from the Lorentz discovery of local time or t^1, which Einstein cleverly interpreted as 'time, pure and simple' through his unique brainwave. Another aspect of time is physical changes 'over time'. Changes that occur 'as time goes by' are not in any way related to time but occur through objects' own chemistry or other conditions like force and accidents. Such incidents happened even when we did not know of time; since then we have learnt to apply time to regulate accessible events as part of rational thought and science.

Actually the sundials give us the wrong impression. They show time as moving physically across the sky; in fact units of time are applied to successive motions of bodies. The seconds do not fly through space; rather we apply them to whatever we encounter whether in motion or not as stressed above. But the application of time takes the form of repetitions of its units: ten minutes means the second has been repeated 600 times. We may not notice it but in logic that is what it is--- that is the metaphysical nature of time. A period of a thousand years starts and ends with a single second. Units of space have been converted to units of duration basically in seconds as our SI of time. The many categories of time (days, weeks, months and so forth) may be culturally useful but of no account in metaphysics. Once the second is defined all units of time are covered since they are multiples and fractions of the second. Understanding the logical status of the second makes the essence and passage of time easy to comprehend. At least some of the legends, bogeys, myths and baseless mysteries of time due to religious beliefs can now be explained logically clearly for those who are capable of rational thought---indeed time can be demystified completely, though we know that not everybody will even agree to look at the evidence! And this brings me to a brief discussion of what role philosophy can play in all that in the next chapter.

For the time being let me explain that the second as our SI of time is also our assurance that reality is there; for the second is a compound of all that we need to survive, all that reality consists of---one second is not just a unit of duration, it is about 20 kilometres of space round the sun; so when it occurs it implies that the sun is there supported by an array of physical and mental objects in existence, that man is also there with his theory of numbers, mathematics and his data of sight. To me it means the mathematicians who thought of the second as our SI of time---since time is the architect of reality[113]---were the real 'Creators' of the world and civilisation not God, as the second is our guarantee that

---

[113] I am not here implying that time is 'given to create'. Rather I think consideration of time, reality, existence, life and the cosmos, when we are finally able to analyse them well, will betray a clue as to the origin of all existence but we're not there yet---and may never get there. Meanwhile random action causing periods of duration (or gaps and the sense of waiting) in millions of objects enable us to have the time lapses that, added to consciousness, lead to creations.

reality is still there even for people who are blind, always reassured by the gravity under the feet; and I repeat that all time is known and used only in units, consisting of either multiples of seconds or fractions thereof.

To know how time is created and used, I usually give the following illustration or analogy, namely, when we say people should not go into the bush at midnight, it means reality has been mentally divided into sub-units---for that is what we use time to do to nature or reality--- and that one of these sub-units is called 'midnight', which is known to be dangerous. That was how it was created and the restriction is what it is used for. But its creation is exactly like counting cycles---one means this and two means that, and so forth. This time system cannot pass through nature physically as it is used for the regulation of nature through the mind only, otherwise the midnight is not physically existing anywhere; it is only a concept formed of darkness. So the really mysterious thing in the world is the human brain which creates and controls everything. The brain makes things easy or difficult in metaphysics, mathematical-physics and the quantum, where we're often astounded even through the brain's own reactions. It really is queer. For example, to begin counting our SI of time after every 20 kilometres or so, amounts to counting cycles to create time in units--- ie, every 20 kilometres is counted as one second on earth, every 20 kilometres is counted as another second on earth, and so on indefinitely, and out of which our time units are constructed through multiplications or fractions.[114] This is counting cycles, especially in a

---

[114] Originally this is not how it happened. At first it was a religious idea that time just happens to be there as divine bounty, and was general and fixed so that a second here is a second everywhere else; but the clever mathematicians kept probing it till they hit on the notion of SI by dividing the orbit of the sun with

repetitive system, and makes the interpretation of time easy because all time units are either multiples or fractions of the second. Give this to any mathematician and the clock and time will be the result as the creation of the human brain, not God. Thus it took Einstein just a few minutes to know that the discovery of local time means time is not divine, in fact much less mysterious than language and speech.

7. THE STATUS OF EARTH TIME IN THE UNIVERSE

What happens on the planet is irrelevant in the cosmos at large. This, of course, is obvious because the universe is rather too hideously vast and complex. In the future I expect the study of time to become scientific, why not? Economics and agriculture have qualified, why not time?[115] It has known attributes, it is a vital subject or entity, we now know how it began or 'begins'---through the intellectual use of points; and therefore what it is made of, namely units of space, not that it just is; it has a quantity for study; we know it will end with the demise of the planet.

---

their own devices to get the basic unit of time which is a moment or a second. They continue to probe it and are now using the cesium atom's unit of oscillation as the basic unit of time used to define the second more accurately. So, in fact, secular time is time by mathematics and that Einstein was far too clever than the rest of us to call local time 'time, pure and simple', and without which I could never have conceived my own theory of time.

[115] Once time is given the full scientific treatment researchers might like to look into what it is. At present they say it just is there; on the contrary I argue that it is obtained through moving from point to point and that without points instants could not exist. But to have a meaning to life here on earth, instants are called 'seconds' or that the basic instant is the second. The seconds are linked to the orbit of the sun so that a certain number would equal to a completed orbit. However the orbit is caused by gravity without which it could not occur. So does it mean gravity plays a role in the having (creation) of time? If so then it means it is involved in our existence too. I personally think gravity keeps our planet at a suitable orbit of the sun for water, vegetation and life to develop.

We know how it is used or how to use it, and we know how it passes by---i.e. by means of replication and not physically through the air. For thousands of years man has been searching the Heavens for grand theories to solve the problem of the passage of time, while staring daily at the seconds repeating themselves to cause this same passage of time---ten minutes means the second has been repeated 600 times. There is no other logical way to account for the passage of time, I am afraid. There is only one second repeated to infinity, and it accounts for all time---the non-interacting moments of Professor Whitehead start (or started) with just one moment and is still the only time there is. Above all, the myths of the divinity of time, its fixed nature and total, absolute cover for the whole universe are gone. It's even foolhardy to believe that one time system could cover such a complex universe!

However, at present time is like the affairs of the human body as against events in interstellar space or the cosmos at large. Man is known to the earth and we can affect many of the natural phenomena on earth. The earth too is known to the cosmos and can affect many events in the cosmos through gravity; so the earth is relevant to the cosmos but not the little creatures crawling on its surface who are infinitely variable and in a continuous state of births and deaths, with infinite variations in uncountable body functions and numerous good and evil ideas floating constantly through their tiny skulls, and so forth---these are unknown to the cosmos and it certainly has no capacity to care about them.[116]

---

[116] This is the perennial problem for mankind: we are completely alone in a world so impersonal and senseless that it has no idea we are even here, while the parent universe is so vast it can scarcely recognise the earth with a magnifying glass, let alone the tiny creatures crawling on its surface---religion too has failed us through the weaknesses of human nature. Thus man's metaphysical loneliness is complete. No wonder some people demand the right to die for one reason or another.

Therefore the minutiae of our existence (including the time we have invented to regulate our lives) cannot be relevant to the cosmos or interstellar space in anyway.

This is my definition of the metaphysical status of man and his time: Day&Night, months, weeks and so forth are all irrelevant. Only the yearly cycle counts because it is caused by gravity and therefore has effects on other bodies. But ordinary, humankind incidents on planets do not count in the cosmos which is so vast that it can only deal with (feel the effects of) massive events. Nothing personal but an incident has to be massive to count or even cause a ripple due to the size of the universe. We are, of course, talking about hundreds of billions of stars, some of them so huge that millions not of people, not even of our earth, but of our sun would find room in them. Can the effects of our mental concepts of time reach them? If not, then how can we accurately estimate their age with our own local time? Who do we think we are just because we have time which the religions claim to be divine?[117] Take, for example, that a person decides to turn round and round perpetually (like the rotations of the earth), or like a fly flipping its tiny wings continuously, why should that mater to the billions of people on earth? Similarly, how could the earth's rotations matter in the cosmos of billions of stars---or how at all could they be noticed and by who or which of the billions of stars in the cosmos? Our time, being conceptual, is applicable to the earth only, exactly as Russell put it: "There is no

---

[117] I have always believed that because of time religious people think there is something important in religion; in fact there is not. Those with the contrary view should tell us where it came from. I agree that those intellectually unfortunate people who cannot live normal life without worship should be allowed to do what will help them, but only as their own personal thing. But the problem is that they don't want to do that; they rather want to slaughter unbelievers or convert them forcibly against their will.

longer a universal time which can be applied without ambiguity to any part of the universe; there are only the various 'proper' times of the various bodies in the universe."[118]

---

[118] From the ABC of Relativity, Ch 5.

PART THREE

COMMENTS

1. THE FOUR KNOWN AGENTS THAT CAUSE THE INTERVALS/ DURATIONS KNOWN AS 'TIME'

We start this section of the book with the Norwegian Question. As hinted in the Preface, a small community in Norway think they have discovered that time does not exist, but they are wrong. It is true that under relativity there is no cosmic time, but there is time; we cannot live without time since we do everything by time. When Bertrand Russell said there is no longer a universal time in his ABC of Relativity, he also asked the question, 'What is measured by the clock?' because, obviously, the clock is giving us time. The explanation of time should begin from this Russell query, and so long as the logical reasoning is flawless, the nature of time revealed will be the eternal truth of the matter; not only that, it will remain the greatest philosophical truth to show the infinite greatness of the Russell/Einstein combination in logical thought.

Meanwhile, some Norwegian villagers have discovered that there is only one day and are excited by that---yet I've been saying there is only one day in many of my books for some time. Of course, the villagers' evidence is more important because it is the physical proof of my supposition, though it cannot make time disappear from the earth. Time is given off by the motions of the earth, therefore so long as we exist, or so long as the earth is there, there will be time. The question is how did we get it if it is not cosmic? What follows is my own fallible attempt to answer this question. We need to repeat the question: what is measured by the clock? One answer is to deduce that it is duration; yet duration is caused in a chain to the basic causative agent which we can never find out. In rational thought as opposed to religious incantations, duration (or during the life of something) can't just happen without cause. So duration is the same thing as 'Being' or existence; and we turn it into sub-units with the application of motion and points. In other words, we reduce existence to manageable units. Thus a unit of time is a piece of existence or reality---what else can it be? Otherwise it cannot be utilized in life. We mechanise that into the clock and apply it to everything, wherever we may be on this planet. Our Norwegian friends can't live without that. Therefore the simple term 'time' is enough for social use while we leave the in-depth analysis to the great philosophers, at least to show that they are useful. But it means time does exist and we can also show why it is not cosmic, general or absolute. This will definitely lead to a new concept of reality, and we owe it to the Russell/Einstein combination. And may I remind the reader that Einstein proved that every planet will have to create its own time system since absolute or general time does not exist, which Russell codified into the phrase, "There is no longer a universal time..."

Yet, naturally, they laughed (as they have always treated me), when I suggested that the Norwegian question is the scientific proof that time is secular as suggested by Bertrand Russell and Albert Einstein---I have discovered that if you mention Einstein's name when you are not writing a scientific article, 'they' will simply laugh at you. However, I dare to invite the so-called academic gurus to ponder this: Norway has always been known as the land of "The midnight sun". But when some Norwegians begin to discard their watches to formalise 'something we have been practising for generations', then scientists and philosophers have got to sit down and think about time, reality and even life to the deepest level. That is what this book is all about, and I find that time is secular, neither general, fixed nor absolute, because there is only one day, only one second and only one year: the daylight is constant and the night is a mere shadow, while the second is either sub-divided or multiplied into the various units of time, and the year ends and we have to restart another year ever so often. And so given the Norwegian question I wonder who is laughing now? In fact, I think the villagers who started this philosophical revolution deserve the highest honours, for the Norwegian question is science, the scientific proof that there is only one day in the universe, and that the nights are mere shadows passing (temporarily) over the constant daylight. This was proposed by me in my philosophy, I agree, but it was practically proved by the observations and behaviours of the Sommaroy villagers in Norway. Also, for 'the mathematically challenged', let me explain that every unit of time consists of a bundle of seconds because minus a single second no unit of time is complete.

May I remind the reader again that due to the discovery of local time cosmic time is no longer tenable, and that the concept of time upon which my theory is based is that time does not run through the universe as Professor Eddington has insisted---**time does not flow. That is what Bertrand Russell meant when he said 'there is no longer a universal time'; Einstein also said, in effect, there are as many [different] times as there are planets.** But time on earth is reality as we see it reduced to smaller or shorter portions. A unit of time is a piece of reality. It means we manufacture (or construct, in Russell's word) time out of our sense of reality. So time and reality constitute one entity but what of duration, since we have varying durations for different units of time?[119] That is

-----

[119] Let me hurry to explain that time is the same reality in regulated units, but

what we need to know, otherwise time is one moment and gone, exactly as defined by the noted mathematician, Professor Whitehead. Time is the duration of that reality. It is **a series of moments, either as the data of sight, touch or feeling---any contact, even in the womb.**[120] **The word 'time' is really a shorthand for the phrase "during the period/event of ..." The word 'during' implies duration, which gives us the sense of time.** The continuity of time arises from counting cycles, like tapping the finger: how many cycles means how many minutes and so forth. In other words, to have time to apply to continuing events we replicate the units of time as created with repetitive motions like the year and similar cycles: we count them, as I said, like tapping the finger, to show the duration of any event.[121] Thus we can have minutes, hours, days and years to apply to any events continuously as part of metaphysical reality; but general, absolute and 'even flowing time' (in the words of Professor Eddington), which is assumed to have been imposed from above, and the same everywhere, is abolished; or not only abolished but

---

because we have to apply points to create the units, the Minkowski theory of a naturally existing 'space equated to time' cannot be logically tenable; but he came close, very close, to the nature of time. Unfortunately, coming close is not helpful as it leads to distortions so much that no scientist knows what time is, except that it just is. Even the Norwegian question shows that to be false. So what is time if not the same as space? Again the reply is that it just is, which is not helpful. I agree Minkowski was very clever, but contend that his supposition was not true. Space time for me means time is based on space and could not exist without it, not that it is one and the same as space.

[120] It is not clear what creates the series of moments Professor Whitehead is referring to; probably they are connected with vision, blinking, the breath or our mental faculties. My personal view is that the series of moments as the definition of time are connected with the multiplicity of objects and the fact that in one picture there can be several objects of varying colour, length, height, volume, duration and so forth.

[121]Thus duration is the real metaphysical sense of time, but the cause of it is not always known, or even humanly traceable.

147

unscientific, with no logical feet to stand on. I hope all that is proved by the arguments above plus the comments below. But the above passage encompasses the whole of what is wrong with our present 'religious' interpretation of time. Yet what is wrong with our time is also what is wrong with our lives, beliefs, civilisation, et al. That is why getting the theory of time right is a serious matter. After more than a hundred years of Einstein, Russell, Whitehead and others, we are now engaged in the serious philosophy of change, not just the philosophy of mere arguments like the Platonic or German Idealism. As usual, I believe the British public, philosophers, and literati are coming to this late, but they will change.

Now, one universal constant of 'Being' is that everything in nature is caused, whether the causative agents and processes are ordinarily experienced by man or not.[122] It is part of the duty of professional thinkers to investigate the causes of things and events as their full-time job. We are in the habit of calling thinkers philosophers; but there are other thinkers who are not philosophers---scientists, mathematicians, logicians, all of these work, in the main, on our problems, trying to solve them for us, and one thing they all agree on is that everything has to have a cause. What we call 'research' is the act of searching for them.

For ever trying to create problems for science, the religions go further to insist that man's life must also have been caused, and as we cause human births, somebody must have set the wheel in motion for giving birth to humankind or we could not have come to exist from thin air. From this they deduce that God must exist, and I would advise people to

---

[122] Even the most extraordinary organ in the universe---the human brain itself--- could not have come into existence unless the elements that created it had the capacity, in combination, to create a brain.

desist from trying to condemn or defeat them in arguments over this because they would rather die than accept any other (scientific, logical) explanation for the being of human life, since by this same doctrine death is only a renewal of life, and yet, despite their pretences, everybody is afraid of death; and of course nobody should try to deny people the right to believe what they will because some people couldn't live without their religious beliefs.[123] Another thing we all agree on is that the causative agents may not be immediately apparent, their effects may delay through inertia or chemical processes, but causes are the agents (creators) of everything, gases, solids, atom, chemistry, et al. Some things happen to cause events just for being there without knowing what they're creating.

Going by the philosophy of causes in nature, so far four natural agents, factors or conditions have been identified as causing time; it does not really matter much how they're described. The reality is that beginning from atoms things are caused by combinations of atoms, whether we as human beings do experience them or not, particularly quantum mechanical causes that are buried deep in objects which were never suspected until the dawn of relativity and QED.

Many people believe that the Einstein Age began with the theories of relativity. In fact, his second greatest discovery was the causes of the photoelectric effect; the first was his theory of time, particularly the discovery that the universe is fragmented and exists by means of different parameters so one system of time cannot be applicable to all of them, as stated above. Wrongly the general theory of relativity is

---

[123] One rule that should never be broken is that once a person is born he or she is entitled to live his or her life to the full---nobody knows how he or she got here so nobody has any right to take life away from anybody.

regarded as his greatest just because of the gravity aspect of it. One can never understand why human beings are so concerned about what happens in interstellar space---it's so far away and not controlled under anybody's conscious direction at all. The British in particular (always sulking and moaning, cursing everybody for the birth of Einstein to come and dethrone Newton), make so much noise about the new theory of gravity.

Ordinarily people think insights about the cosmos are more valuable than insights about the little pieces of matter that actually make-up the mass of our bodies. But to discover that the world of visible matter is actually controlled by the movements of the invisible quanta, is my number one discovery of all time, (the Nobel Committee also agreed), followed by the idea that time itself is created, in the words of Russell, "constructed" by man not imposed by God, and bound to vary from place to place. Gravity comes third in my estimation.

It may sound like fantasy (even treachery to some philosophers), but I think the religions believe the problem of time can never be solved, and that after Darwin time is what makes them believe still that God exists and that he created us. It's no use citing biophysics and all the rest of it. They say time and life go together. We know life because we are living it. We can't know the nature of time though we are using it. It's therefore open to a variety of interpretations. Choosing the logical or scientific interpretations does not preclude others from opting for the religious view. We have to remember that when we use physical cycles, like the yearly cycle, to track time, what we are doing in fact is showing or recording the rate of the passage of mere physical cycles and never the time itself----whatever it may be.

The reason, of course, is that the physical cycles are not time; and we don't even know how long the year is in terms of duration in the mind. They give us just how many times (orbits) the earth has recorded. We count these cycles and call them 'years' or whatever. So that for mankind, ten orbits of the sun is ten years; and one would have deteriorating physique to show for it. Is that time? Certainly not. We count them as the rate of time but they are mere physical cycles; so it means we count the physical cycles as the rate of the passage of ten years. There is no way that these physical cycles could be the real nature of time, only how it is passing by. The division of the year into months and so forth means nothing---we still cannot define one unit of time. For instance, how long is one second in logic? One minute? One hour? They don't have meaning because we can never know what duration means; counting the years is just a matter of counting cycles, tapping the finger is the same thing. It proves nothing.[124] Perhaps we age as the passage of time. Ah, ten years will make everybody age in some way---more or less. So can we regard ageing as the basis of time? But how much do we age as an indication of the passage of one year? There is no clear landmark, and also it varies from person to person.

In fact, ageing is chemical and we actually base our time calculations on ageing without realising it; time is based on ageing not the other way round. What influences us to call chemical processes time is the regular day and night system, the passage of which we regard as synonymous with the passage of time and the ageing of our bodies---namely, due to

---

[124] It's from such arguments some writers insist that time does not exist. My reply is that it is not an honourable stance for philosophers to dismiss something they use daily as a myth. Time is not only in real existence but is oppressively unavoidable.

religious beliefs we (mistakenly) suppose that the days are passing by due to the natural passage of time in one direction towards the Day of Judgement etc., and as they do so we age 'over time'. This is the universal human belief but completely wrong in both chemistry (meaning science) and logic; it makes the life of those writing logical interpretations of time extremely difficult. For a start, there are no 'regular' days in astronomy or nature at all; there is only one constant day. You can divide it as much as you like yet it is still only one constant day. The rotations of the earth are completely irrelevant---something like human speech or footsteps, they do not count in the interpretation of phenomena because they are not metaphysical entities; like the grunts of cows, or tears of men and women, they do not count. So it means our notions of the passage of time and therefore all traditional ideas about time are wrongly conceived. Every statement that includes the phrase 'turning back time' should be changed. We cannot turn time back; it is a chemical process in the head, no one can turn it back.

One can see that the religions are no fools. I believe totally in secular time; but those who do not cannot be dismissed as easily as that. And since the religions have money and power they want to keep their privileges with as much mysticism and doubts about nature as possible. That's what they are doing. It's not fair, but there you are. Many things in life are unfair to the losers; but you can't blame wealthy and powerful people for trying to preserve their advantages in life. Life is so nasty that anybody who has any advantages to help him or her live comfortably should seek to preserve it. After all, self preservation is the brain's most powerful, oppressive force on people. You can't avoid it and live.

In any case, stated below are the purely secular factors or conditions that can cause the sense of duration we use cyclical motions to divide into the physical units of time (strictly speaking it is a confidence tricks.)[125] We use these for reckoning time for general use either individually or in various combinations. Obviously time is a construction; as such we use the parameters we assign to nature or recognise in our environments to construct it, and therefore bound to be different on other planets. But time anywhere else in the universe will have been 'constructed' the same way—by the uses of either one or many of the following agents. The Russellian notion that time is a construction seems the most logical explanation of time under relativity. Not that time cannot exist under the Einstein proposal as some writers claim---that was an expression of intellectual laziness---but that it is constructed by the human brain. So it's the brain we have to decipher atom by atom, if we can! Not only about time, why is it able to probe the entire cosmos? The whole idea of being in existence, of being human, centres on the brain's extraordinary incisive powers. What is the purpose anyway, since we are so ephemeral and infinitesimal? For me it shows the infinite generation and regeneration, death and rebirth round and round indefinitely. Pythagoras was nearly right, except that the same material does not last; atoms disintegrate. Otherwise what is the meaning of regeneration? So, for me, death is sadly the end. Traced from the nature of time, this is philosophy at its deepest level.

Let me point out that even the theory of evolution, as scientific as it seems, may be questionable under secular time. Scientists have kept it quite, but the problem is this: divine time was planned but secular time is not---we construct it out of elements found on earth, meaning it did

---

[125] They work out of necessity.

not exist for whoever planned evolution, yet all planning requires time. So evolution may be just one of those consequences of gigantic cosmic events; but bearing in mind that what we know of the universe is the product of the human brain, was it imposed or part of us? If it was imposed, by whom? If it is part of us since we feed it with our blood, then how can we, as worthless and vulnerable as we are, probe the vast cosmos with the puny organs of our own locally constructed time?

This is how important I find time to be---even then we can only know how it's passing by and never what it really is. Perhaps the greatest mystery in all nature is time, and this is how it is caused from my point of view:-.

a) Chemistry (physical and organic);
b) Quantum mechanics;
c) Motion;
d) Inertia/force.

I will now give brief comments about how these agents can cause what we experience as time or a period of waiting, always by means of the brain, the real mystery in all nature.

First, Chemistry: the most obvious example of chemistry or chemical processes causing time as a period of waiting is human gestation or any gestation at all; though human pregnancy is what concerns us most. The normal period is nine months. However nine months constitute the time for the growth of a fertilised egg (the foetus) to the point of birth or a normal baby---ready to face the world, with what luck nobody can tell! We call the waiting period 'time' but in chemistry it is a long process of growth, the conversion of matter from one form to another.

Next, the quantum debate: quantum mechanics is not perceptible, but they cause a period of waiting, though most likely to be very short---it is still time.

The third factor is motion. Motion, obviously, can be seen as time. It takes time to move from A to B. For instance when a sportsman throws a ball into the field of play, it takes time to travel to reach the players in the field. We will normally analyse this in terms of dynamics, kinematics or kinetics, in a word mathematically, but in ordinary perceptions the effect is known as 'time' carrying an immense philosophical or religious baggage. The important point is that it means what we know as time is caused by physical events, elements or forces and not imposed by God.

Motion of any kind can show time going; but it is never the real time in force. Real time, as I have argued, is never known; only its rate of passage is recorded by the clock together with its psychological effects as the sense of duration. In other words, <u>if real time is unknown, then the cycles we use to reckon time constitute the time in so far as we are concerned.</u> This is what is taking mankind so long to grasp since the abolition of universal time. Mankind did not have to worry about time when it was supposed to come from above, but now the first Literary Agent kind enough to write back said, "I am sorry but not being a professional mathematician or philosopher... I am afraid I cannot figure out why this is important or what it's about..." Brilliant! I am also afraid that it is important because (in the absence of a universal time), scientific progress (chemistry, medicine, physics, electricity and all that we need for the proper and safe control of our environment) depend on the proper understanding of time consistent with the world we live in---the sun, daylight, weather etc. The problem is that he holds the money, the power and the publicity levers and he does not understand me, or the fact that we need to seek the real nature of things so as to be able to control or manipulate them to our advantage. Even Einstein admitted that he was able to complete the special theory of relativity a few weeks after he got his insight that the Lorentz t^1 is 'time, pure and simply'. Lorentz also said he thought he was unable to discover special relativity because he failed to take his discovery about time as important---and dare I mention that the practical benefits of relativity are too numerous to mention? Without time we can never tract dangerous asteroids.

What Lorentz didn't know was that the genius in Albert Einstein was just looking for the 'fact' he discovered about time to change the world, from one of living by the grace of divine time to the one of living by our own construction of local time, the philosophical implications of which we are still struggling to elucidate. Otherwise time dilation is not important as the whole of time cannot dilate from the workings of a single clock.

To go back to our story about the clock, the units produced by the clock are obtained elsewhere and programmed into it for reproduction. The ticking of any clock is not original; it's meticulously conditioned to do it precisely in the form that is done. The units of time are deliberately created (and mathematically divided) to accord with a full orbit of the sun, since we then have to start the orbit of another year. Often people equate motion to the rate of time. Such erroneous statements about the metaphysics of time are all over the place. Every time you mention time you're making a metaphysical statement about the world. It may seem familiar but you could not form such ideas if you're ignorant of the planet's motions. And we do it all the time. Even as I write a newspaper report (The Times, 16/9/13) is repeating the mistake.[126] They wrote:

"Flies and small children may have something in common: the ability to slow down time...by seeing the world in slow motion..." This equates motion to time but that is totally wrong. Time, of course, can be based on motion; and every motion can be seen as 'time going'. The problem is how much time? For this reason not every motion is metaphysical time. It is some time going but how much in real time? That can only come from the true nature of time, and that is obtained from the breakdown of the yearly cycle before being programmed into the clock for reproduction in specific units to accord with a full orbit of the sun. It's only when you understand this will you come anywhere near the metaphysical nature of time.

Otherwise if any motion is time then the faster the motion the faster the time and vice versa. That could not give us a stable environment; for all activities are controlled by time. This is additional evidence that motion (any motion) is not time per se, and all statements to the effect that time is mere motion should be regarded as mistaken. For they go on to claim that gravity 'slows' time or speeds it up.[127] We can only do that by

---

[126] Frankly I am not surprised that my books are not read. Everybody thinks he or she knows what time is---even flies can slow it down, really?

[127] When Einstein was questioned about the Twins Paradox, he replied that one or the other might have been affected by acceleration. We may give the same reply to the question about gravity and time. However, it all depends on how the time is defined or explained in logic. I am rather concerned that the same writers who pose this question also suggest that perhaps time might not exist at all. If time does not exist then how can it run faster or slower in certain situations? As they are asking the same sort of questions about Entropy and time they may soon come up with questions about time and politics, time and agriculture, time and rain, etc. Nobody can answer the numerous legends about time or any other thing. That does not prevent a logical explanation of time or anything else. If my explanation of the passage of time is true, then whether time is this or that here and there is immaterial; we can always say we do not yet understand some

affecting the motions of the earth, but that cannot affect other scientific theories. For instance, relativity is obviously true and did not falter simple because Einstein subsequently failed to explain the Twins Paradox. As Bertrand Russell pointed out, science cannot explain everything but it can destroy the world; it means it has got hold of some of the vital knowledge of the universe, and must be obeyed (with appropriate safeguards) on pain of death.

Real time, again, is unknown. What we do is use mere cyclical motion to show how much time is passing. We are dealing with something almost like quantum mechanics: we cannot demonstrate the entire range of quantum action within matter or atoms that result in chemical changes at the visual level. Yet through the theory of QED we know that they do occur and that life depends on them. We couldn't exist without them. Time, of course, is slightly different because we've lived without mechanised time before----when we're very, very primitive---and could probably do so again. (In fact, the historical study of time fails to reveal why it seems so mysterious, because it evolved and we can trace its evolution to its very beginning. But then there is religion, fear of death and the yearning to come back after death together with wild expectations of time travel. Why the human imagination latched on so fiercely to time travel backward and forward beats my understanding. Now they're even saying the Minkowski theory of 4-D geometry makes time travel 'a scientific possibility' through 'curved space-time'.)

I would combine the effects of motion and inertia into one causative agent. Motion and inertia combine to cause activities in stellar bodies. If a planet is moving to or being sucked into a black hole, it could take

---

aspects of reality and move on not give up..

centuries, or at the very least long enough time for us to get married, raise children, grow old and die; even for several generations to do the same thing; or fight wars; conquer other nations; establish civilizations, lose them, and start all over again! This is the reason evolution can happen in human calculations. Otherwise evolution, as a cosmic event, would require time and planning which do not exist in the universe. Therefore even evolution is freakish---more than a million degrees down the scale of cosmic events, and therefore unrecognised. With evolution man thought he had discovered one of the basic events in the universe, but it is not so. Logically we're mistaken. You couldn't plan evolution without time, but there is no systematic time in the universe.

All of these either singly or in combinations can cause time interpreted as a period of waiting---not as 'time allowed' by 'The Creator'. It is an understandable mistake caused by philosophical ignorance to claim that the periods caused by all or any of these alone or in combination do not constitute time like the time assumed to have been bestowed by the mythical Creator of the discredited religions---which means all of them.

To digress a little, here is my philosophy as regards the act of worshipping deities. The prescriptions (liturgy, Gospels, prayers, incantations, etc.) are revealed by ordinary people. They call them revelations. I call them dreams. In life you have to get education to survive, even as a child you have to be taught how to use the toilet to survive. So education is crucial. But the basic thing we learn in education is how to reason; how, for example, to know that certain objects that resemble food should not be put in the mouth---how to form ideas about things. First we identify things, observe them, establish what they really are as far as possible (e.g. dogs are different from tigers, so that we would not go and play with tigers), and act accordingly, or according to tradition and create safety to yourself and others. This process is called logical reasoning. Without it we would be fantasising about the world and the things in it, but won't last long because there are so many dangers from snakes, tigers, people and inanimate objects that can very easily put an immediate end to life. Given this precarious conditions of human life on earth (where we are forced to live without knowing why or how we got here), it is unwise to rely on somebody's mere dreams as the prescription to regulate the course of human life, especially en-mass. We know it is unwise because it has led to numerous accidents and ruin. Science gradually evolved after centuries of trial and error to help us identify the nature of things; it teaches us how to deal with them in safety. No religion can do that and therefore all the religions are dangerous---but there is fear in some people that unless they hear something sweet about life and after life they could not live, as they're taught that life is at the mercy of God. Unfortunately delightful religious sermons are not associated with safety and progress; it could even be the reverse, as some people dream that the Gods require human sacrifice and come to take you away. We have learnt through bitter experience that some people are basically evil and infiltrate the religions just to wreak havoc with their fellow human beings' lives; no matter that they utter agreeable sermons---it is a deliberate ploy to seduce people and ruin their lives. For this reason worship of the Gods dreamt up by others is unwise and very, very dangerous. But what about our own dreams, then? Again, dreams are outside logic and therefore unreliable; it is better to stick to what has been rationally examined before you trust your life to it. The unexamined life, we're told, is not worth living.

We are talking here about secular time: after Lorentz and Einstein found that absolute, general or universal time does not exist and running all through the cosmos and the same everywhere, from the past to the

160

present and the future, it became immediately necessary to investigate how our own time began; and the line of thought given above has proved irresistible. It is logically the most convincing reason for the existence of time.[128] It should be emphasised strongly that time was not only thought to be fixed but that it was the same everywhere. That is the reason we think Einstein, being so clever, had a brain wave that if it is changeable then it could not be the same everywhere and so on. This may sound elementary, even tautological, but every little thing about time takes a genius to clarify. Nothing is more mysterious or linguistically intractable.

THE ORDER OF TIME SEEN AS A MATTER OF ARITHMETIC

Why is time so oppressive that even school children know that time waits for nobody? If they have to go to school at ten, ten it will be, and nobody could delay the time itself---they could delay themselves but not the time.[129] This is the main reason which makes time seem to be an

---

[128] I believe the bogey of past, present and future was spawned by the study of history as a story from the past, giving rise to the present and carrying on to the future. People think of the process as one of 'time marching on'. It is precisely as they describe it but it is not time that is marching on---it is the events. Suppose you borrow money from your bank, it is the money the bank manager will chase all the way from the past to your present circumstance and your future earnings, not the time; that was merely the position of the earth round the sun when you borrowed the money. Bank managers are notoriously careless about time, not the money. They'll gladly extend the time of the loan if it will bring them more of your money in the future!

[129] The reason (or the science) is this: every second is 'a space interval' constructed with points out of the space traversed by the earth round the sun in one orbit, as mirrored by the clock's hand, but nobody can delay, stop or slow down the motions of the earth; so the time moves on unstoppably. This 'order of time' is what causes the oppression of it. Repeat the process continuously and

imposition by the Almighty. The religions are very fond of it. We even have a saying in all languages throughout the world that time waits for nobody. The reason, I have found, is this: once we have divided the orbit of the sun into units of time running strictly to coincide with a completed orbit (where 31,536,000 seconds=one year), the time units became oppressive—no one could change them without skewing the alignment with the orbit of the earth round the sun and get into danger, especially during the night in ancient times. That is the reason time follows a strict order, for the units have been carefully worked out to accord with the motions of the earth. The basic cause is the order of arithmetic. Time is no longer as mysterious as it used to be barely a hundred years ago. The oppression and passage of time added to its ordered nature and sequences were the three insoluble aspects of time, making it seem divine; but all these can now be explained logically and conclusively, if we begin from the premise that it is constructed (as Russell deduced).

The human mind craves order and cause, so we think of the order and direction of time—but by who, or imposed by whom? We've just been liberated and freed from the bondage of time's absolutism only to be confronted with its restrictive order and direction. I think it's all the fault of the brain. The brain requires order and direction due to the way it's put together. We imagine that several compounds cobbled together accidentally to create the brain.[130] With each component searching for 'complement', the probing tendency in the brain was established as part

---

you get a perpetual time system to cover any planet or world---as part of the logic of time in the universe.

[130] Recently scientists have created a human brain in a laboratory by the same methods.

of its basic structure. This may be mere speculation, but to me it sounds credible.

Even then, there is some tenuous evidence for this because we can see how it grows in the foetus as the host (the woman's womb) supplied it with the necessary compounds and chemistry, and when complete, begins to take over the foetus and cause its growth to the point of birth; after which it is fed externally to direct the body to the point of death. Originally it must have grown compound by compound. The process probably was that one compound and another joined up accidentally. Then the quest for 'complements', the need for order and the sense of waiting resulted, caused by what I can only describe as 'chemical hunger'. Most of what is written here cannot be proved; but I think it may come close to what actually happened. For something must have happened to cause the generation of the brain out of inanimate matter, elements or compounds. This is an attempt by ordinary human beings to trace the physical origins of the human brain. It's not rocket science, as they say. The important thing is that it's not religion or fantasy either, for the life itself grew the same way: elements formed the original, egg or sperm to create the basic primeval amoeba that replicated till it grew to be a sustainable organism. It is also evident that the brain's demands on the human body (including forcing us to have and endure the sense of waiting which we know as time), are built into the brain's structure; and we imagine that it could have come from the protracted, chemical processing over many centuries involved in the brain's creation out of the elements. This word 'element' is used instead of saying atoms, but it is to be understood that its ultimate constituents are believed to be atoms. This, of course, is the language of science. It is conceded that not everybody believes in science; what is evident, however, is that nobody can live without science. Some of us abide by it on pain of death.

The Origin of Secular Time

Otherwise it's not illegal to believe in anything that causes no harm to others.

To continue with our story of how the brain probably evolved, we can imagine that when the required compound for complement arrived, more chemical tentacles were created---for the brain is more than a million times more complex than the biggest computer ever created by man---more materials were needed for a complete organism to reach an independent and sustainable whole. Hunger, the sense of waiting, yearning, cravings, the need for reason to explain causes and the need for completion to calm the increasing number of chemical tentacles requiring 'soothing' (roughly, it is thought), caused the brain to come to exist with these tendencies built into it. One of them is time, or the sense of waiting (waiting for the require complements), which internally we know as duration. Another is the craving for order and causes. They do not exist in the universe outside the human mind, that is the reason the cosmos seems so chaotic---I believe it's not governed by time or order but by accident and chance. Even the world is similarly chaotic, unless a human being is involved or is controlling events.

Otherwise by whose order or direction is the putative arrow of time following? It may very well be that the order and direction of time are misleading concepts, precisely the manner we get the years and the centuries. Time does not seem to move physically only mathematically; you can easily alter ten year to twenty on paper or through mathematics; but the physical aspects of time are external and come from the motions of the earth. Thus altering ages (or the figures of time) on paper is useless; the physics will always win by skewing the human estimates. After that human ingenuity took over to create concepts of time in the mind to accord with the motions of the earth, which may be regarded as the birth or the cause of the birth of our whole concept of time. Since the earth's motions are repetitive, the units of time spawned by them are also digital, and we applied arithmetic to them. Thus, unlike our primitive ancestors' practice of keeping time with charcoal marks on the wall to indicate the number of days, weeks, and months gone by, as the passage of time, modern man has the use of theory to simplify things for him---after that the mystery-makers took over, beginning with Pythagoras.

Anyway, because of all this, we now know that time is produced in units and the units multiply for it to advance or move on. Is that still disputed, with the example of the yearly cycles being so clearly demonstrated? But it's essentially a matter of arithmetic: if the year is one unit of time then it replicates to become two, there, four, five, six and so forth all the way to the centuries. The yearly cycle is not moving past the signposts of the years all the way to the centuries; this is obviously not the case. Rather the year is repeated over and over again indefinitely. Is that still disputed by the mystery (or mischief) makers in the universities? It has the merit of confirming Russell's idea that we actually do construct our own time, probably through the brain's instinctive cravings for order; so the order of time is in the mind not in the world out there. Sentience, a theory of numbers, the ability to count and points are required to fulfil this order. Hence space is involved, since the use of points implies space for the creation of theoretical time based on mathematics or arithmetic. We can still call it 'space-time' but only in the sense that time is or can only be created in association with space; and its passage, too, is the same as its ordered progression---i.e. through the procession of its units, a matter of arithmetic. The years increase in numbers to pass by, without any direction. For one thing we've been able to establish clearly is that things are created continually in the universe through the accidental combinations of elements, compounds, atoms, et al. Even the brain created itself and dies to prove that it is not a permanent entity, just a passing thing produced through accidental causes.

All the same, some writers have made the order of time the pivot of their own interpretation of time, even though the definition of time as the 'irreversible general passage of existence' was meant to refer to a time system imposed on us to run generally in one direction through the entire universe and the same everywhere, with the condition that we are all moving in tandem to the Day of Judgement as the end of time--- the biggest folly in the human mind's suppositions, incredibly, illogical and totally without foundation, just part of the Christians' meaningless fables that we have wisely rejected for over a century ago.

After its rejection the order of time should have been seen as a purely logical matter easily resolved with arithmetic.[131] Let me explain. The phrase and particularly the word 'irreversible' mislead people into thinking that we are all moving irreversibly with time (plants, animals, entropy, vegetation, the seas, rivers and streams, et al) to a predetermined end or destination. This has spawned numerous legends, theories and beliefs mistakenly; yet the order of time merely means the arithmetical order or progression of time's units---the years, for instance; exactly like counting the years from one to a century. And one year is also pared down to the seconds, that should reach a certain number (counted progressively in arithmetic), to coincide with a

---

[131] Time is arithmetical, so it is oppressive; it imposes order, the order of arithmetic: from one you just have to go to 2, then 3 and so forth---oppressively---that is why time is 'oppressively' unavoidable; for since the arithmetic is based on the point-divisions of the earth's motions (being the space traversed by the earth), and the earth never stops, time's progression is oppressively inescapable.

complete orbit of the sun---and start again. Because of the restart, the number of units counted has to be exactly correct; therefore time cannot wait for anybody. It really cannot do so physically. The phrase does not mean an irreversible passage of time that we cannot interfere with, but irreversibly leading us to doom. This is what scientists have magnified into the theory of entropy's irresistible march to the death of all activity. The theory of time has accumulated thousands of myths, fables, fantasies, legends, lies, religious beliefs and even mischief. I am quite sure several sacrifices have been carried out with human beings about time to placate the Gods. Yet the order of time which some writers regard as an insoluble problem implying divine influence is nothing more than the progressive counting of the units of time, after all that is how we get the centuries.

The order of time and the supposed irreversible passage of time should have been eliminated from the debate as soon as we realised that time is discrete as intervals or time units between points; for obviously we do not all move as such:[132] the Heaven is nowhere; God is supposed to be dead; and existence is not even uniform, neither do we all move in one direction. In quantum theory the direction of motion is not even known. Some things are stationary, others are moving in reverse, and others are moving in any way they prefer. Even in the solar system presumably controlled by the sun's gravitational attractions, not all the planets move

---

[132] The order of time and the progression of the units of time from one year to a century amount to the same idea, solving the problem of the advance or passage of time. There is no problem of the order of time to be resolved, if the time is discrete and not running through the cosmos in some kind of continuous thread from the past to the present and on to the future. In our ignorant past we thought the order of time was also fixed. But once time is seen as discrete the order becomes a matter of counting the units.

in one direction. In any case, once we found that time is variable, it was unwise to insist that we are under the command of one kind of time moving in one kind of direction to a solitary mythical destination. We now know that time, existence and motion are all variable. This is the situation Einstein discovered, namely physics is for the planets because there is only chaos among the stars (too big and complex for anything else). Time also belongs to the planets. For the cosmos to have regulated time somebody must live among the stars who has the brains to apprehend cyclical motions that could be used to reckon time sequences.

In spite of all this, everybody comes to the study of time with his or her own agenda without reference to logical truth because religion and ancient traditions have so conditioned our minds that we all think we know what time is. Einstein alone showed (he did not just say it; he proved it by experiments) that we are all wrong, and Bertrand Russell not only agreed with him but said that his theory of time was, perhaps, his greatest achievement. For me there is no doubt about it. General relativity is not Einstein's greatest achievement. It's not even his second. It's his third after time and the quantum theory. I am of the opinion that what we find in interstellar space is of secondary importance to what we find here on earth; for even if we discover a body on a collision course it is what we find on earth that could be used to neutralise it. In most cases interstellar knowledge is mere intellectual pastime for selfish, psychological satisfaction. Something much more like an ordinary labour of love. The near star mentioned above may explode and destroy not only life on earth but the earth itself. But what can cosmologist do about it? ---absolutely nothing; as I have said, except to give us sufficient warning to go and join Richard Branson in his special plane, but even then where to? Could there be any permanently safe place in the cosmos?---to want to live forever is certainly not a rational idea. We are born to die, but why? What for?

I know that human vanity is bigger than the sky, but obviously the universe is just too big and complex for us to worry about putting anything right up there. The causes of its nature and activities can hardly be more important than the price of bread here on earth. No matter what we do or believe no course of action by human beings (as insignificant as we are) can make any difference (other than, perhaps, diverting or destroying asteroids on a collision course.)

Although not a believer, I sympathise (somewhat) with believers about the purpose of human existence---what for? Man is so insignificant. Even the planet itself is just a tiny dot soon to end up in a black hole and burn out of existence; yet human intelligence is so far-reaching, so inquisitive, so hungry for knowledge of the cosmos (probing, probing, and probing), to no avail. For all his insight about the cosmos, Einstein died, decomposed and disappeared out of existence altogether. But his insights and discoveries about the world we live in are in daily use for the benefit of mankind. They certainly are more important than knowing about black holes. So while the human brain's creations can last, we the creators, the bearers of the brains, could die easily. Men are so fragile. The religious people are not that stupid---there is a real problem with human life on earth; there is human yearning for salvation; we just do not want to come into the world to die in misery---what is the point of that? The religions want to claw at the tiniest straw that could be interpreted as giving meaning to the senseless thing called life: the best philosophy for the worst reasons. But the alternative, as we know, is sheer emptiness, misery and early death for no sensible purpose whatsoever. Thus, despite my basic irreligious beliefs, I still think

religious tolerance is the beginning of genuine humanitarian wisdom---
"Remember your humanity and forget the rest", was the last advice
Bertrand Russell gave us. And Russell, for the ignorant, was not just
another human being. He was the world's greatest philosopher at the
time, and most likely as clever as Aristotle.

However, as the result of the new Einstein theory of time, we now know
that there is truly no longer a universal time as Bertrand Russell put it in
his book ABC of Relativity; so we have got to search for the mechanism
of time or how we get our time anyway--- or what we call time. Atomic
time is included in earth time, as part of the time we have created with
the earth's motions. It is often assumed by some religious scientists that
atomic time constitutes a cast-iron proof that time exists in the cosmos
and can be measured in many different ways. In fact, atomic time is not
different from earth time. The cycles, pulses or oscillations are merely
shorter than the long orbit of the sun, and, in any case, they have always
to be related to the second to make sense. For this reason atomic time is
still part of earth time. One can even tap the finger. It is the same thing--
-something we can count as the rate of the passage of time is all we can
have for the reckoning of time. It is through sheer hard work and
amazing human ingenuity (mostly by mathematicians), that we have
'constructed' what we call time to guide our activities; and since this
time is based on the earth's conditions, not all of which are conducive
(or compatible) to living without sensible controls, our time is strictly
tuned to show us the safe periods and areas of the world's conditions
and environments we can negotiate in safety.

Otherwise there is no time in the cosmos at large. Our time is unique;
there is no doubt about it. For it is created with the unique parameters
of the earth. If there is natural time behind the parameters we simply

cannot know it, because that is not what we know as 'our time'. What we call time, or our time, is 'constructed' from the parameters as their effects only---counting physical cycles as 'years' is not time. Perhaps they are the effects caused by natural time behind the parameters, but we simply do not know.

As already mentioned, Einstein divided the universe into two. They are the metrics of general relativity where there is no tolerable conditions for human lives, and the metrics of special relativity where life is feasible. In this home of ours there can be time, as we suppose that in similar homes elsewhere in the cosmos time will be thinkable. Where there is life, there will be time. That is part of the logic of time in the universe. Every time system can only be based on physical parameters; and they are all different one from another. Some or most of these parameters are present in all the segments of the universe; otherwise there is no time in the universe. To have time you've got to have the intelligence to construct one out of the components of the relevant parameters.

The conundrum is this: on the one hand, we think there is something called time naturally moving on by means of the factors or agents mentioned above; but if true, what then is moving this time on? On the other, it would appear that, like the brain emerging from nowhere and seizing control of everything in sight till its own demise, human ingenuity has created what we call time out of the natural features (or parameters) we find in the universe. It appears these parameters or features, being mere physical materials, would know nothing about time as the sense of duration in our minds.

The Origin of Secular Time

Being the greatest philosopher of the period under discussion, Russell asked the most important question about time, enough to redeem philosophers' reputation after their condemnation by Karl Popper. As already noted, when Lorentz and Einstein showed that absolute, fixed or general time permeating the whole cosmos (and the same everywhere) does not exist, Russell asked, what then is measured by the clock? Frankly, apart from the orbits of the sun, there is nothing (unless you can tap your fingers continually as the rates of the passage of time). Hence the thought of secular time.

A careful analysis of the Russell question gives a perfectly logical explanation of all aspects of time as a secular entity 'constructed' for use on this inertial frame, and even then only capable of showing how much time is passing and never what it is---provided one can ignore the billion or so myths about time.[133] We use cyclical or regular motions to give us time--but they are physical, so they can only show how much physical cycles are passing (have passed or will come to pass, e.g. as 'years'), and we use them as time periods to plan all activities: ten hours means it is time to do so-and-so for so much hours, etc. What the real time 'is' we can never find out, only how many cycles (being the years) of it have passed or are passing. As I have said, my guess is that time is a combination of chemistry and motion (especially repetitive motions) and sentience; none of these on its own can be mechanised into a clock as time, but in combination they can give one 'a period of waiting' (especially in chemistry), which is time. Sentience is required because somebody must be there to set the points and count the orbits of the

---

[133] This answer, though very theoretical and academic is not much different from counting the days as the passage of time---or even tapping the finger to indicate the passage of the seconds.

sun as years or there will be no years and seconds derived as fractions of the year. Until we were wise enough to do so, man had no time and lived like a beast of the forests. Just look at the story of the evolution of the clock since we came down from the trees. In brief, time is life plus activity.

One implication of all these contradictions is that the phrase 'Space-time' may be sensible, succinct and cute to some scientists and those mathematicians who want to reject the 3+1 formula for representing physical reality in space, but it's not actually true of the physical world, unless it means time can only be gained through the application of points to space, so that we get time units (or time intervals) as relation between points, like the years, not in the sense that space and time constitute one entity---just to avoid use of the 3+1 formula--- so that as space curved in general relativity it would take time with it, as 'curved space-time' for you to meet your grandparents even before they were married, just to justify religious sermons about time travel being 'a scientific possibility'.[134]

---

[134] I am trying to reveal what I know or suspect of the innermost yearnings of many pure mathematicians. We know that they tend to be mystical in their thoughts; but it is not right to dress-up their secret yearnings as objective truth. That is not how we acquired the dependable knowledge to go to the moon and back, or cure some of the dangerous diseases that plague human life. It's not fair. We spend billions to keep these mathematicians in the style of life conducive with calm and serious mental exertions only for them to make spurious claims that time travel is a scientific possibility just to satisfy their own secret mental cravings. All those who believe that time travel is possible are free to leave the earth in the next hour---an hour is long enough for them to pack their stuffs. Professor Eddington's advice was that in an experimental science we have to discover properties not assign them. The fact that he gave this advice in the Introduction of his Mathematical Theory of Relativity has added poignancy.

The Origin of Secular Time

It is obvious that the universe has no time as an oppressively unavoidable order of action, as we have on earth that is why we get problems with the quantum and other sub-atomic particles.[135] Neils Bohr said whoever is not shocked by the quantum theory has not understood it. That statement should be turned on its head, namely whoever is shocked by the quantum's strange behaviour has not understood it---he or she does not realise that the quantum alone (without conscious direction as in LASER) is not subject to time or what we call human time, 'constructed', out of matter after the quantum came to be in existence. I would advice that we look at the quantum carefully. From what we know of its nature, we understand that it would have been there (in a strange sort of existence we can never imagine) long before their interactions caused objects to come to be in existence. It can be in two places at the same time because it is the most natural piece of matter behaving without the influence of time. It was there before we invented our time and therefore cannot know how to obey that time. It is outside order in nature, it is not directed by anything; the human mind's notions of order and time sequences do not apply to it. You would be shocked by its strange behaviour because you can only judge it with a shallow mind that came to exist after the quantum, and is therefore unknown or recognised by it. This is looking deep (speculatively) into matter to the quantum level, so deep that physics cannot include it---but that is how physics itself came to exist and yet it works to the extent of having the capacity to destroy all life on earth.

---

[135] These existed billions of years before visual objects were actually caused by their own interactions to come into being. How, therefore, could they copy the actions of visual objects? They don't know them! This, of course, is speculation, but the whole of solid science grew out of such speculations.

As the most original matter and the smallest bit of matter that can exist, the quantum is not subject to any of the human concepts about order, time, motion and chemistry as we know them. The quantum existed before the regular cycles we use for time, order, motion, chemistry. This is how the (quantum) Nobel Prize winner, Professor Richard Feynman put the same idea: "The word 'quantum' refers to this peculiar aspect of nature that goes against common sense"---exactly.[136] It belongs to a universe before the common sense came to exist and is therefore not subject to any of its notions. Common sense refers to common objects of the perceptible universe. The quantum is not part of this universe. It's the most original and basic matter whose many and varied interactions have created phenomena as we learn from QED; it therefore does not know how to behave to suit us as we are part of that phenomena to which it does not belong.

We can construct time out of the phenomena in our experience, that's our peculiar luck or curse. It all depends on how you look at it.[137] All the elements for this act of construction are everywhere in the universe but not as time (to the universe). They are rather events occurring haphazardly under a variety of forces: gravity, space, inertia, motion, heat, chemistry, without conscious control. The religions are right about one thing: the process of human creation requires intelligence. Where did it come from? They claim to know that, but cannot prove how they know it. References to the scriptures are what annoy scientists most.

---

[136] See the Introduction to his famous book QED, Princeton, 1985.

[137] To some people life is a blessing; to others it is a curse and regret that nothing in life is worth living for because it won't last and may even end in tears---love for instance. Nobody can enjoy life from birth to death without regrets.

The Origin of Secular Time

Again, our instincts expect the quantum to respect time or behave according to time sequences---but the universe has no time. The parameters we use to reckon time are purely accidental events. We know they occur, but cannot think of how anybody could have organised them in the manner we arrange things on earth. As we have come to realise, there seems to be no direction in the universe, no purpose and no logical sequences. Existence is existence; it's just there. There can be no doubt that life emerged as the result of chemical accidents, but, like everything else, it just happened for no purpose at all; and while it seems to us to have been long in existence, in the cosmos at large our period of existence is just a flash, and the earth itself just an infinitesimal dot, not worth bothering about. We see nobody there to worry about it either, unless and until we set things and events in earth-time.

Ah, but the universe has no time! Earth time is just that, namely a time system we have created for ourselves on this planet and applicable to this planet alone. That's the conundrum, and I for one finds it enormously interesting just pondering it, usually alone, as my religion. Some people weep over life's problems. My advice is to try and find them interesting as events occurring without cosmic control or significance, and yet so vast and complex that pondering them is itself rewarding as an intoxicating spiritual solace. It's not true that it will make you mad---it'll rather cure your madness!

Nobody is there to infect you with the germs of madness. In nature things happen haphazardly. There are some limited logical sequences like something cold fleeing something hot or ice melting at certain temperatures, but no streams of logical sequences (such as we human beings can construct out of this huge and complex admixture of accidental events), by means of the human mind including the consideration of time---the most essential thing besides life.[138] So it appears that outside a human head time does not exist. According to Russell we 'construct' it ourselves. I agree with Russell absolutely. I do not believe that time does not exist on earth because we are using time everyday; but it certainly cannot be defined logically.

It's obvious that time does exist on earth; the problem is how to define it. So serious is this problem that we have come to the conclusion that its logical definition amounts to just how it is passing by. So we think we can only know how it is passing by and never what it is. But in the universe at large, although we can spot some of the parameters we use for time on earth (everywhere), no one is constructing time sequences out there through the use of these elements, not from our point of view anyway. All notions of time are carried from the earth to apply to the cosmos in breach of the Einstein theory of frames.

CONDITIONING THE HUMAN MIND FOR TIME

I regard this as the appropriate juncture to explain the nature and importance of time before we carry on. I agree there are past, present and future in the world or the universe, of course. Einstein probably meant they're human concepts for our convenience, not basic and permanent episodes to be allowed to influence theory; that's my understanding of what he said: the past is obviously what has occurred most of which, but not all, would exist only in memory; the present is what is on-going, and the future is what we would expect to be the consequences of the present and past put together. Logically all this is

---

[138] Whoever solves the problem of time will come closest to knowing the nature of human life of which the mind is the most mysterious, for it's the mind that creates time.

beyond dispute. However, as related to the interpretation of phenomena they do not count; they are not relevant at all, and can only lead to confusion and wrong ideas about things. For they are based on the concept of 'passing days and nights', which constitute man's notion of the passage of time. Yet there are no passing days and nights in actual, physical reality in the whole of the universe. There is only one constant day anywhere, and it never moves, dims nor closes down even for a second. If it does life will be extinguished. The days and nights are temporary blips caused by the earth's rotations but they do not count because the earth is so insignificant. There are so many billions of gigantic bodies (stars, suns, moons, etc.), that if we were to use any one body's motions to interpret phenomena, we could end up changing our ideas so regularly that stable life would not be attainable. What the earth leads or misleads us to suppose cannot be used to interpret or influence the cosmos. Above all nothing is passing through the universe in some kind of a thread, least of all time, which consists of separate units (like the years) and passes by through the succession of its units, again, like the years. Hence all other units derived as fractions of the year are also separate and individual and succeed one another. This is the most logical explanation of time we have for scientific use, deduced from just four thinkers in all history, namely Einstein, Leibniz, Bertrand Russell and Professor A.N. Whitehead.

But still speculating about time, I think if it is true that every inertial body has to have its own time, then there simply is no time in any part of the universe until you have created your own time and conditioned your mind to its nature, and Einstein made it clear that we can only do so in inertial bodies, not in general relativity. Conditioning our minds with our time is bound to lead to changes not naturally existing in other parts of the universe. The elements we need to construct our time are not available in general relativity; and if they are not available even in general relativity then the other world the quantum came from would not have them either. As noted, Einstein divided the cosmos into two[139]: one is where you can have time by sustained regularities (or 'constructed' logical sequences lasting long enough for the human mind to use for its creations), and the other is where, because of the strong gravity, you cannot even see anything anywhere at all to have the necessary regularities of motion to use for time. For time means from when to when, from one point to another---the year, for instance. Part of the problem in physics come from the improper understanding of Einstein's ideas, for all time is based on regular or repetitive cycles---the year, for instance, and since it is the cycles we count as the rate of the passage of time (like the years) we can never know the true nature of time only how it is passing by. Through our mathematical ingenuity, we've learnt to use cycles to provide units of time. I call this the quantification of time. These are what we use as time: years, hours, minutes and so forth. They merely indicate how time is passing by, obviating the need for complicated theories about how time passes through nature. For a start, our time is not even passing by, or passing through nature. It is discrete and proceeds unit by unit---year by year, minute by minute and so on. But in my experience it seems mankind doesn't want to know about discrete time. Man is so enthralled with time passing through nature to the Day of Judgement, and start all over again due to the transmigration of souls. It seems nobody wants to die if he'd not come back to life after his holiday up there! Or live in another

---

[139] Perhaps it should be three: Inertial Frames, General Relativity and the strange world of sub-atomic matter. Similarly, the Two Postulates of special relativity should be three with the addition of time---the parameters that can be used to construct time sequences should be there.

world up there. Thus it's easy to get some religious people to commit acts of terrorism in suicidal attacks.

Let me emphasise again that what we call time is only how it is passing by---the years for instance, pared down to the seconds. But Einstein is so misunderstood that many writers insist that we need the Minkowski formula to understand relativity; yet the special theory of relativity that concerns us most on this planet had nothing to do with the general theory of relativity and Einstein's use of the four-dimensional continuum of Minkowski in the equations of general relativity. Rather it merely amounts to incorporating time into space to form one entity so as to dispense with the 3+1 formula. There are suggestions about the usefulness of this procedure, but originally it formed no part of the theory of special relativity. In other words, the Einstein theory was complete without Minkowski in so far as special relativity is concerned. The Minkowski theory did not improve special relativity---it's already complete and critically acclaimed. All suggestions to the contrary is evidence of ignorance; for it is obvious to me that only those working closely with relativity understand it without using the Minkowski formula for equating space to time in one equation, yet it was never successful. Professor Eddington was right when he struggled to recall the name of the putative third professor who understood the theory in the initial stages. Looking back, I believe it was himself, Bertrand Russell and Professor Whitehead. I don't think Minkowski ever recognised that 4-D geometry is no help to relativity, meaning that he therefore did not really understand relativity. The culprit is always religion; even the best thinkers cannot free themselves of the thought that time is eternal and imposed by Providence. Otherwise it is difficult to see how any intelligent man can accept that time is incorporated in a universal entity like geometry. It is an ephemeral entity created for the convenience of ephemeral mankind.

Let us look at time in practical terms. Our parents agonise about what will happen to us when they are dead, because they know life would continue as it had happened to them when their own parents died. So when they're gone, we would be there. It means the world will carry on when we are also gone. If the world carries on it means geometry is carrying on. Yet our time will not carry on when there is no one there to set the points for the cyclical units we call 'one year', and out of which all time is derived as fraction. In plain words, our time will end with the earth's demise. How therefore can this ephemeral time be part of geometry which, of course, is eternal? The theory should stop at deriving time from space with points, meaning the time could not exist without the space---and therefore it is 'space-time'.

Presently as I see it, the religious people want to resurrect the bogey of past, present and future to prove Einstein wrong. They claim that the syndrome causes the flow of the story of history; that history is what the study of the past to the present tells us. Yet past, present and future can be perfectly logically explained as uneducated fiction[140], so Einstein was right: the past is obviously memory; the present is now, carrying the past as historical baggage with it (you never leave your problems or wealth behind you, do you?); and the future is mere speculation. History is the march of these events not time; and the events are still marching on as the continuing story of life. Thus the past is not still existing anywhere to be revisited---it is here with us as the consequences of what happened in the past! History is not seen as time running through nature from the

---

[140] Or the illiterate man's understanding of history.

past to the present and so forth. Our time, as the year shows, is discrete. Discrete time cannot march throughout history (or the cosmos) as people like to believe. Only events have antecedents and consequences. The times are added as the times of occurrence. It is the events that mater. Many of the religious-based mysteries of time can also be resolved. I have published ten monographs about post-relativity time explaining all these issues, but nobody is showing any interest as people continue to chase their emotional thrills from worthless books and gadgets. I fear that true culture is dying slowly due to the aggressive onslaught of the electronic strangulation.

If time is not permeating the cosmos and moving from the past to the present and going on to the future then there is no quandary. You can challenge this theory of time, but by the yearly cycle we should know that time is discrete, from year to year, repeated over and over again for all the centuries; there is only one year in all the universe, repeated to carry on as years; and every unit of time, too, is derived from the year together with its astronomical features. If time is like a thread passing through nature, then it is reasonable to search for theories to account for how it is passing. But we have a time system that is repeated to continue. The year is only one; to have two years we repeat it; to have a thousand we go round the sun a thousand times. Surely everybody can understand how this time passes by as something in procession---units of time following each other? There's no need for a theory to explain how time passes by. This is all we call time. Every unit of time is a fraction of the year as divided with points or astronomical features. Time units have no independent existence; they exist only as fractions of the year no matter how they are derived. The mathematicians have done a good job about this; but there is no mystery; it's plain common sense.

We've nothing else for the reckoning of time except mythologies or counting the days as the passage of time in a primitive manner without theories. I must repeat that there are no years at all in nature existing as something we can just pluck out of the sky and apply to events. Let me repeat again that there is only one year and all other units of time are fractions of the year. To have more years we simply go round the sun again and again; that is what we know as the passage of time, or it constitutes the passage of time---namely the units of time in procession. This idea solves at once the fearful conundrum of the passage of time.

We hear so much from writers about the passage of time, but nobody has ever been able to define time. You have to define something before knowing how it behaves. Once time is defined as relation between points, like the year (and that is not in dispute because that is how we get the year, and the year is time), it becomes something proceeding unit by unit or intervals of time in procession causing the continuity of time---like the year increasing in numbers all the way to the centuries.[141] After all, human notions of time come from the year and daily rotations of the earth, or the day and night system. This time can only proceed unit by unit; so the passage of time is seen as the procession of time units---the year for instance, worth repeating a thousand times to defeat the stubborn doubters and critics of rational thought. And as these units are all passing it means all we can ever know of time is how it is passing

---

[141] The really original thought about the Einstein theory of time is the Lorentz discovery that time varies from place to place. After that the inferences are straightforward---for that is precisely how we get the year---and if time varies then it is not general, fixed or absolute, covering the whole universe and the same everywhere. It is an interesting example of correct definitions leading to permanent solutions of intricate problems. Even the Time Dilation idea it was based on was not correct, yet it helped! Time cannot dilate.

by and theories of the passage of time are redundant, even humbug when it's incorporated in religious sermons and the Day of Judgement mythology.

Even then (strictly speaking) going round the sun is not time. Rather it gives us a long duration of reality we can sub-divide into manageable units of reality called 'time' or time units. We use it to show how much time is passing and never the true nature of time; for going round the sun is a physical activity, yet we count them as years because we have nothing else to indicate how much time is passing. The years replicate to become centuries; or they increase in numbers to pass by. Using the yearly cycle to know how time is passing is not time that is passing. Time does not pass, only the units do; but the units are mere physical cycles. "A time system", as Professor A.N. Whitehead has said, "is a sequence of non-interacting moments"----year after year after year, or as pared down to the seconds and the other fractions of the year, and he made this observation in his book entitled The Principle of Relativity---I am not afraid of these repetitions because this is a theory about time; a subject so contentious that new ideas about time ought to be hammered home or fail to convince the critics, no matter how true they may actually be. On the other hand if my critics have got the message they should tell me so that I can relax or even retire altogether! After all, I'm over eighty and not in the best of health.

Everybody on earth agrees that time is mysterious, yet the scientific study of time after Einstein is becoming logically consistent and even delectable as logical solutions delight us all; the human mind craves logical thought; and if we reason logically from the premise that there is no longer a universal or cosmic time, and that the basic unit of our time is not even definable, so we simply do not know how long the year is and therefore all measures of cosmic time are flawed,[142] then time becomes easy to understand. We normally say human beings age 'over time', and the year is regarded as the best yardstick of age and ageing. The truth is that the year has nothing to do with ageing. We age through chemistry and metabolism but they take time to occur. Everything takes time; that does not mean time is causing them, the underlying causes are always there if we look hard enough. Time takes the credit or blame out of ignorance simply because it is always there as the chemical, accidental and physical causes of events, these we choose to call 'periods of time', again out of ignorance.

But we can't tell this to the cosmologists wasting the taxpayer's money on their pet projects---like smashing atoms. They say it brings technological, economic and medical spin-offs---yes indeed they do, but researchers could achieve the same results over time without wasting billions. How many billions did Bill Gates spend before hitting on his ideas for lucrative ventures?

Brains are what we need. In my opinion the CERN is a complete waste of money. And they're not even sure of their theory. They announced recently that if c has been breached then they might have to re-examine the concept of 4-D geometry to see if it is really true. The point is that

---

[142] Time dilation, the twin paradox, clock paradox, and so forth---none of these is even half as important as showing that time is neither fixed nor absolute but changeable, thus, at once removing it from the realms of religious thought to us down here on the ground as a secular entity. Tracing what it is, is proving difficult, but at least we know that it originated from this earth and limited to it. We know how it all began; we can trace it back to when we're mere apes and had no sense of time.

the Minkowski ict equation is flawed for being based on the imaginary time coordinate, i, not because of the status of c.

Being a grumpy and infirm great grandfather, I am too old to fear of what could happen to my career; others are probably silent because they have good reason to fear the powers that be! Scientists chasing research funds are as ruthless as the Mafia, probably more so. But there is a lesson here. Science is different from any other calling. It is so open that any theory in any branch of science that is not true cannot be hidden for long.

Scientific mysticism has always been part of the problem of time's definition, but now, thanks to Eddington, everybody can be sure that there is no such monster called 'Time Zero' from whence time is supposed to have began and running all through the universe ever since from the past to the present and the future till God calls a halt to the whole damn thing on the day of judgement; a childish fable forged on us as the true and most profound philosophy of existence. To my mind this is intellectually shameful. Religious believers may be gullible, but the rest of us are not that childish to believe an infantile fable like that.

By the same token, 'curved space-time' by which time travel is said to be 'a scientific possibility' by the very people awarding the Eddington Gold Medal to their clever fellows is totally untrue. But sure, physics has got to put its house in order. And the physicists must start with no illusions about our time having any influence in the universe, for the whole earth is only a tiny, tiny, tiny little infinitesimal dot in the milky way, let alone the entire universe; and the psychological-time constructed by the 'worms of the dust' crawling on its surface (as the poets describe us), can hardly influence the universe at large, although I know that greedy publishers wanting to cash in about time travel based on human gullibility, will continue to publish such books (as mentioned above), implying that we and our time can have some kind of influence on the cosmos as a whole. There ought to be a law against the spread of such falsehoods that make people less not more rational. The universe is certainly mysterious; but time is not so strange any more, since we know that at some stage in our lives we simple had no time because we lived like apes on trees. Since we came down we have tried many things for telling the time, the most rational of which is the earth year, pared down to the seconds and the atomic pulses---that's the most logical explanation of time possible and we owe it to Albert Einstein alone. I will now try to sketch what time was before Einstein and what it became after the great man.

THE NATURE OF TIME BEFORE AND AFTER EINSTEIN

I estimate in my own little way that very few scholars who understand relativity are aware that Einstein made an important discovery about time, yet he did. "That is the gist of the June paper's kinematic sections" (see his Biography by Abraham Pais, Ch. 7 Opp. Cit.), and it is more important than anything he did in science---which is saying a lot. But it is part of the reason they called him philosopher/scientist. However, because scientists despise philosophy and yet time is supposed to be a subject in philosophy, not much is heard about Einstein and time. It is a surprise to me that scientists fail to realise that nobody could alter aspects of physical theory without having to think like a philosopher[143]. At the highest stage the two subjects are almost one and the same thing. In what follows, I will try to share what I know of the scandal with the reader, because, to my way of thinking, physics will eventually encounter so many difficulties that scientists will rush back to the Einstein theory of time---namely, it is limited to a frame or planet, implying that the universe itself has no time running through it, and whatever happens in interstellar space occurs out of chance, chemistry or accidents. The creation of Man, for instance, must also have been an accident. The implication is that time is a human creation.[144] Given the mysterious nature of time, this is a large claim, and, if true, then the nature of reality comes into serious question...[145] Whether we can do anything about it or not, we just want to know.

---

[143] Science is not just proof, proof, and proof, even in mathematics. At some stage, as Goedel has pointed out, some equations are going to have to be based on unproven premises, and yet they'd work even though philosophical, so philosophers are not that useless.

[144] About thirty years ago I wrote a little book entitled, Why Time is Not a Natural Phenomenon. I had to pay more than two thousand pounds to see it in print. I was given fifteen copies and that was the end of it.

Before Einstein---the philosopher--- time was supposed to be general and absolute, such that any unit of time here is the same everywhere else, a creation ultimately attributed to God. After Einstein we see it as a secular entity that is limited to a frame and discrete because we can reckon its essence and passage only with repetitive cycles, **which is the process that makes it discrete.**[146] **The actual motion of the earth round the sun is not the time. Motion on its own is not time. Rather the orbit has been mathematically divided to create recurrent points (exactly like recurrent cycles.) That is what gives us our units of time; and it proves that Bertrand Russell was right to call time 'a construction'. The movements of bodies (especially repetitive motions) give us the cycles we use for reckoning time so as to say, for instance, an event was there for so many cycles, or seconds, minutes etc. My argument is that in the absence of this theory of time no one can account logically either for the nature, provenance or passage of time. I have been saying this for more than fifty years, and I will continue to say it---as I keep repeating- -- till the good Lord calls me home since mankind is too devout to appreciate rational thought, but rather prefers the fanciful explanation of religious sects.**

Obviously these 'cycles' can only give us discrete units of time, year after year after year, indefinitely. Thus we have years forever. We then pare 'the year' down to fractions ending in the seconds and the atomic oscillations based on the second. In general, our time is based on the orbit of the sun; the units of time as the levers of mechanics are based on the moment or second as our SI of time otherwise we couldn't live, that is why all time units are either multiples or fractions of the moment

---

[145] Is the United Kingdom the only country in the world where such ideas stir no emotions at all? Not even the courtesy of a reply? It makes me think that if Einstein was British he couldn't get a job even as a labourer.

[146] To the religious sceptic, this is using logic to analyse an entity that, in the absence of logic, is assumed in a thousand different mythological ways that are useless to scientific thought.

or second. That is how important time happens to be. But it did not exist before, and will not remain after man, though the elements used to create time will still be there in senseless nature. We pay far too much attention to nature when the important thing is to use it, for, by itself, it is permanently senseless. On earth man is King. Man cannot sustain civilisation without the SI of time which is what provides all units of time by multiplications or fractions, and therefore, by implication, all time. But time that is based on a standard unit is bound to be discrete, being produced with points unit by unit. From that it is even easy to explain why this time cannot flow through the universe, hence Professor Eddington's angry retort about 'the even flow of time'.

Also, as some writers have observed, Einstein's theory of time arose from experimental results and therefore not open to doubt. So the scientific answer to the question 'where does time come from?' is what I have sketched briefly above. Generally, mankind has not become aware of this and continues to treat time as the same as existence---this is the reason the Minkowski proposal is so popular; it touches a yearning in people. Nevertheless, let us look at time before and after Einstein in rational thought, for there is always rational thought; some people are born with a logical mind, others just the opposite.

But before we do that, and in order to placate the haughty scientists who despise philosophy with the ferocity of viper attacks, let me show why philosophy is important as the absence of philosophical oversight for time is leading to contradictory statements from scientists and, in my opinion, also causing confusion in physics when it comes to discussions regarding reality. Time controls everything, therefore it is basic to all things, and everything we do, know and think about---including the real nature of reality or ultimate reality.

Our time, since Einstein, has been understood to rely on points and therefore not universal (see Bertrand Russell's ABC of Relativity, Ch. 4-5). Thus we have been able to create the SI of time. The problem is that a time system (as Professor A.N. Whitehead argued in his book, The Principle of Relativity, Cam. 1930), that is based on points must be human in origin, and Russell concluded that it is bound to be time that is 'relation between points'. (Note that so far not a single scientist has been able to consider this matter to the depths of the philosophers---except Professor Stanley Eddington, who said angrily that any contrary views amounts to making 'meaningless noises', as mentioned before---in fact, I quote him in all my books; so infuriated was he, yet I think he was absolutely right.) As he said, before Einstein's researches, we all believed in the concept of time flowing through the universe. After Einstein, the idea is no longer tenable. But time is subject to legends more than any other subject; so it is no surprise that people still regard it as if Einstein had never existed. However writers and scientists have a duty to be more accurate in their references to time.

Then in 1999, the editor of NATURE, no doubt in his noble role as one of the pillars of the scientific establishment in the world, wrote his book, What Remains to Be Discovered. In the Introduction he wrote: "Uniquely among the laws of physics, the second law [of thermodynamics] specifies the direction in which systems change with the passage of time; it defines what has been called the 'arrow of time'..." To a philosopher, this is worst than the religious view of time because time produced with points is discrete time. It cannot run through nature in the form of a stream; discrete time cannot flow, has no history and can have no arrows---yet that is the time we have discovered through scientific experiments. Furthermore, many processes in science depend on time, but this time is human in origin. What does that tell us about

191

the nature of ultimate reality? To refuse to discuss this matter out of the hatred of philosophers and philosophy is an act of intellectual madness. Yet the Institute of Physics rejected my book even without reading it---- all credit to the Bodleian Library but OUP is too important to answer the likes of me. In any case, I am allergic to moral and pastoral theology for which they have created a professorship.

But when will publishers, mathematicians and scientists realise that there is no 'passage of time' to cause entropy? They have to consider the matter afresh for their present position is intellectually worst than the garbage from the religions. Entropy will happen at some time in the future, no matter what we do or do not do; but because there is always time, when it happens it will be at a certain point in time. The time is not in the air to infuse all material reality with time; it is created with the use of points and therefore is human in origin. As such it can't be anything other than discrete else we couldn't create the SI of time. To write √-1.ct, and conclude that because of that equation s=ct (to mean space is equal to time) is a fanciful hoax unworthy of scientists, and I will continue to say so until the good Lord calls me home. Sadly scientists do not see these things because they never read the philosophers. Not all of them are good, but since Russell, many of them have become mathematically competent or literate. Ignoring all philosophy will never cease to hurt scientists. For example, if there is no time in the universe and everything happens there by way of chance, then there can be no mechanics in the cosmos only random events---therefore entropy can and will be caused by any events. To see the universe through the eyes of man is a mistake.

TIME BEFORE EINSTEIN

Generally speaking, few writers have studied time seriously with a view to suggesting cogent theories about its nature before Einstein.[147] Even those who made such attempts, like Henry Bergson, were guilty of assuming that it is just there.[148] That we find it in existence, and that is that. Even the careful, logical thinkers fared no better. They made it look synonymous with motion or 'Being', eventually calling it the "irreversible passage of existence". Yet existence is not one; it is multitudinous and individual. This means at any moment billions of movements are taking place: some (like leaves, just waving in the air), moving up and down, sideways, forwards and backwards, tumbling, limping, dancing, rolling, crawling. Moreover every individual is uniquely separate with his or her own perspectives---no two persons, as Einstein showed with his analysis of simultaneity, perceive one event identically---space and time coordinates are involved. The multitudes of people perceive the world differently. Above all, existence is not altogether passing in tandem. In

---

[147] Kant also tried. No doubt his numerous followers who believe he's invincible would be looking for his name. He tried his hands at everything in philosophy, and that is the reason he achieved very little of lasting value. Kant always lumped several topics together in sweeping statements, linking physics, physical reality, astronomy, cosmology, logic, linguistics, metaphysics and even psychology in a mesh of contradictory doctrines. As always, Bertrand Russell put it best in his History of Western Philosophy: "To explain Kant's theory of space and time clearly is not easy, because the theory itself is not clear"---I would say the same thing about everything he wrote.

[148] Henri Bergson was better than Kant on the subject of time; a great French philosopher in his day, he's one of the very few brave thinkers to write a book about time, though he spoiled it with thoughts about 'Freewill'---the two don't mix very well. However he regarded space and time as completely independent of each other. They still are.

quantum physics directions are not even fixed; what may be irreversible to you could be the opposite to somebody else looking at the same event from another angle. It is necessary to mention that Einstein also failed to decide how time is created, and stressed only that it is neither fixed nor absolute, adding the most revolutionary idea that it originated from this inertial frame and that there could be as many times as there are inertial frames, thus laying the foundation of secular time.

To discuss the thinkers I commend, I have to stress that, contrary to the opinions of some scientists and particularly the pure mathematicians, philosophy is not as time-wasting as is generally put about; it is so serious that it shares with theoretical physics the ultimate attempt to formulate credible theories to account for our understanding of the nature of the external world, or the cosmos as a whole, including psychology and cosmology, or the mind and matter ---what happens in our heads and the entire universe of sentient beings. There have been great and valuable contributions to human welfare and material progress from philosophers. Apart from what one may describe as 'the scientific thinkers' like Pascal, Archimedes, Aristotle, Pythagoras and the rest, only a few philosophers have actually made valuable contributions to human progress, but they are there. I am thinking of writers like Plato (with all his faults he discovered the idea that we can never perceive reality as it really is by his most original, first-class supposition called 'The Simile of the Cave'. He also promoted philosophy and taught Aristotle.) Rene Descartes' merits included the Coordinate Geometry, upon which relativity is built, and of course, The Cogito. Our own Bertrand Russell's many merits include the new logical theory of 'Denoting'. Henri Bergson speculated that space and time are separate entities. If we had listened to him we would not have wasted so much time and energy over the Minkowski fiction that they are one entity---

which, I believe, is still causing distortions in physics and relativity. Professor A.N. Whitehead and Bertrand Russell also discovered that the world of sense is 'a construction not an inference', which liberated scientific thought to an enormous degree and still bearing fruit. For example, the photons 'construct' images; this idea can be seen as much more rational than supposing that we perceive pre-existing images created by God, which we are only able to see by the grace of God as they are invoked by the mind out of thin air as Plato proposed in his theory of Ideas, and which has been used to justify the existence of God by the religions. In monumental efforts, Russell and Whitehead again tried to derive all mathematics and arithmetic from logical premises, and even though it's judge to be unsuccessful, many logicians believe that it inspired Kurt Gödel's Incompleteness Theorem, otherwise known as The Theorem or The Proof. However, as we have seen, Gödel later ruined his contribution with suggestions that time travel (a notorious religious myth) may be 'scientifically possible' and that even Einstein agreed with him!

The Cartesian Dualism and his notion of God as the Absolute Perfect Being are not mentioned because they are religious notions and religion causes wars and does not advance philosophical knowledge. All the religions borrow from two basic ideas. Due to the fear of death as the end of human life, they borrow from the Pythagorean 'Transmigration of Souls'; and also from the Rene Descartes dichotomy between the soul and the body of man, or Cartesian Dualism as it's called. We don't want any of that because it promotes religion and religion causes wars. Human life is bedevil by a fantastic paradox: the majority of mankind believe that they need to worship and couldn't live without it, quite apart from the hope that it could make death just a transition to a better life; yet the organisations they establish for this worship cause wars and sectarian conflicts that make peaceful life on this planet so hazardous that even if life after death were possible it wouldn't be worth having.

Returning to our main subject, time, as I have said, had many different meanings for different thinkers before Einstein. I identify four such meanings all of which, due to the importance of time, have become the focus of mass consensus, following, or even civilizations and worship.

First, we have the clever and profound, world-shaking, intellectually brilliant interpretation of time proclaimed by the Irish Prelate, James Ussher that God created the universe at exactly noon on AD 4004, implying that time was already running from the so-called Time Zero, marching on to give us the story of history. For generally history was only understood as the march of time. Let us call this religious view 'Act of Creation'. Even school boys can see that it has not much intelligence to commend it. Yet the majority of mankind, numbering billions seriously believe that this is how the world came to be in existence, and worship anything they believe to have 'Created' it. A religious interpretation of time that nevertheless links it close to life as all the logical analysis suggests. There can never be a final, definitive theory of time satisfactory to all mankind. Properly, we should all live with what we believe and forget about the rest---except that time is a powerful agent of causes in science.

Next, Isaac Newton. Newton believed in absolute space and absolute time, and since he was very great in science, everybody able to think followed the great man's definition of time, and it ruled scientific thought until Einstein. Even still now many scientists speak about time as if they think it's running all through the universe like a thread—or that it's general. To be honest, I really consider those who claim against the facts that time does not exist more intellectually respectable than their opponents, mostly scientists and philosophers, who insist that it just is, whatever they mean by that cryptic assertion. In the words of one book reviewer, "Time does not flow, it just is", a biblical language expressing a biblical myth. For if that is so then what is the use of the earth-year---why should we try so hard to find a logical explanation that ends in the existence of the year? I rather accept Bertrand Russell's theory that it is constructed as "relation between points". Thus sentience is required, because somebody must be there to set the points for the yearly cycle; which means we are sent back to the beginning as the religious people claim that only God or a Supreme Being with infinite powers could guarantee that somebody with the necessary intelligence would be there to do that. So, in the end, we have to realise that nothing in life is easy to explain without an ultimate attribution to a deity, least of all time! Even though we know God does not exist out there--- at least Bishop Robinson told us so in his book 'Honest To God'.

The third movement is the Lorentz/Einstein denial of absolute time and absolute space, demonstrated (or proved) with scientific experiments. Soon after that we got the fourth interpreter of time, being his own mathematical interpretation of the Einstein notion of time, called the Minkowski formula for space-time continuum, or 4-D geometry for short. Its brevity belies its momentous effects on scientists. Because it was supposed to be scientific, or the mathematical interpretation of the Lorentz/Einstein experiments, practically all scientists refer to every time and every space as 'space-time', meaning that space has been equated to time---that the two entities have become one. And yet Minkowski could not define time. Also the very great scientist and mathematician who confirmed the general theory of relativity, as we have noted,

The Origin of Secular Time

Professor Sir Arthur Eddington, the founder of astrophysics, described the theory as arbitrary and fictitious.

I have therefore rejected the Minkowski formula as logically untenable, since it is based on imaginary time coordinates. The 4004 Creation of the universe is too dumb to think about, and the Newtonian absolute time is abolished by Einstein. My rejection of the Minkowski theory is not mathematical in the sense that it is not written in mathematical symbols but in words. However it is somewhat mathematical in the sense that it is a logical objection and all mathematical statements have to have a logical premise—and it is his premise I am challenging. I am simply pointing out that in the absence of a universal time Minkowski had to show where his imaginary time was coming from; never mind that it's imaginary, but from where? Also how can anything imaginary be relevant in discussions about time?

So we are left with the rational consideration of time Einstein proposed, namely that time is derived from your local space---points together with the ability to count and sentience are required, exactly the way we obtain the year, which is the basic time unit on this planet. For the philosopher, the problem is not that time does not exist since we are using it daily in all sorts of activities, but how we get it, how we 'construct' our time, because cosmic time is abolished due to the discovery of local time---if I have to repeat it. This is the current logical situation about time overall---that is, scientific, philosophical and practical. Now let us look at the specifics, or how some writers have considered the matter.

With regard to the view that time does not exist at all, my understanding of the thesis of Professor Yourgrau's book mentioned above is that the legacy of Gödel and Einstein to the effect that time cannot exist under relativity has been scandalously neglected or forgotten by the world. Additionally he says, when that is given its proper due, time travel becomes a scientific possibility.[149] But this cannot be true because the very idea that there can be such a thing as 'The Logic of Time', as proposed here, is derived from Einstein's researches; this logic makes time necessarily discrete, so that we count individual years to get the centuries and our own numerical ages. The fractions of the year too (the seconds, hours, minutes and days) are also separate and individual units of time. Obviously, there is no chain in time; every unit is uniquely separate---Professor Whitehead's 'non-interacting moments'. That is what is meant by discrete time. However, discrete time, such as we have here on earth, makes suggestions of time travel laughable. Yet the sad fact is that most of the recent writers on time seem to be only interested in theories that make time travel appear to be "a scientific possibility": "Gödel, the union of Einstein and Kafka, had for the first time in human history proved, from the equations of relativity, that time travel was not a philosopher's fantasy but a scientific possibility."[150] Kurt Gödel bears most of the blame. He asserted that his discussions with Einstein had convinced him that time travel is feasible. I concede that scientifically that could be so if time is naturally equated to space; but so long as they

---

[149] It is difficult to see how time travel could be possible if time does not exist. It is also not an honourable intellectual stance to argue that something human beings cannot even avoid does not exist. It may be intellectually difficult to define but that we know something we call 'time' is indisputable. But then we clearly have to understand that intellectuals have a bounding duty to explain the reasons why and how an indefinable entity is in daily use. That is the situation with time, God and life. Of these three basic conundrums time alone seems (not only in existence) but positively decipherable.

[150] From A World Without Time, Opp. Cit., Ch 1.

are separate entities they cannot 'curve' together as to suggest that time travel could be possible. The Minkowski attempt to merge them with mathematics was not successful---even he himself admitted that before his proposals space and time were separate entities. And yet, as discussed below, in nature we human beings can never make two phenomena one or one entity two with mathematics---see App. II.

Thus I wish to assure the reader that there are absolutely no such (logically valid) equations in relativity that could even remotely make time travel seem plausible. In special relativity, there is none at all, and yet that is what concerns us most, since there is nowhere anywhere in general relativity to be capable of giving time there to anybody. The so-called interpreters of general relativity carry earth time there in clear breach of the Einstein theory of frames. What happened is that mathematicians coerced Einstein to incorporate the Minkowski formula for 4-D geometry into the field equations of general relativity, which was very easy to do. Yet the Minkowski formula is logically flawed and therefore completely unacceptable as the basis for altering physical reality to one of four-dimensional space, in fact, it bears no relations at all to physical reality and looks beautiful only in his own mathematics, which is considered arbitrary.

Let me point out that what the mathematicians wanted was a new formula to replace the 3+1 system (as they do now, even though they know that it is not logically valid), so that they could write one equation to represent time, space and matter as 's=ct...' It was an extraordinary demand to address to inanimate nature, and would be funny, something like a silly prank, if it were not so serious.[151] Man is an orphan because

---

[151] In all life (from the beginning to the present), we have come to realise that

he does not know whence he came. He is a clever orphan/ape on a lump of rock; and for such a creature to demand something from nature by way of his own mathematics is nonsense. Yet they went ahead, and Minkowski obliged. However, Professor Yourgrau himself has quoted David Hilbert as saying Einstein did not believe in the concept of 4-D geometry or four-dimensional space. Therefore, for me, space and time remain separate entities, especially on any inertial frame like the earth, as Einstein made them in special relativity. I hope the pure mathematicians will stop playing their usual games where relativity is concerned.

In all history (if we ignore the German Idealists, as I am trying my best to do), time has been regarded as the same as existence, The age-old definition of time as 'the irreversible passage of existence'. In effect, equates motion with time.[152] In addition, since nobody knew (or still has any idea) how life came to be, no one worried about the nature of time which is closely associated with it. Hence the mathematicians chose to refer to time as 'just is', meaning it just happens to be there in nature; and so time and life were lumped together as constituting the eternal

---

logically and realistically, addressing demands to inanimate matter is a sheer waste of time---like all prayers. It may help through psychology but not through the alterations of material reality. Physics means material movement or manipulation, without that material changes on the prayers of another material entity (man) is sheer nonsense.

[152] I have explained above why motion cannot be equated to time, although it can be used to reckon time either singly (like the earth's motions) or in combination with other factors, such as chemistry. So I see the attempted equation of motion to time as an interesting paradox of contradictions: if motion is equated to time and yet the motion takes place in space how can the time be equated to space? Above all, time can only be had through the application of points to space! We're seeking to bind our thoughts with stultifying contradictions.

mystery on earth. The genius (unique insight) was the suggestion that it does not even exist until you have invented it out of the parameters in your environment. This is the Einsteinian notion that Kurt Gödel misinterpreted. For time does exist on earth; the fact, however, is that it did not exist before we came down from the trees. It means we created it; and since its creation it has come to dominate our minds and everything we do, like water (or language). Nobody is born demanding water in order to live. But once we taste it we realise that life cannot go on without it.

The mathematicians used astronomy (being the natural features of the world, as perceived) to construct clocks, and that was it. We simply used the time provided by the clock makers. This time was taken as general and absolute; it was assumed to have originated from divine sources; and since a centrally imposed time system could not be different in different places, one second here was supposed to be one second everywhere else.[153] Then Einstein burst on the scene with his new idea of time, showing that the old idea was religiously imposed and not really true---obviously time is neither general, fixed nor absolute.

Given the supreme importance of time in human affairs, the irrational view of time (before Einstein), influenced all life and all ideas including religion and historical narratives adversely, spawning mythologies many of which are still with us today, with their own consequences, some of which are even detrimental to human existence. One of these is that history is the march of time---marching since the 'Dawn of Time'--- instead of recognising that history has been the story of how life has

---

[153] Nowadays, I guess no educated person will believe that one second here is the same everywhere. But in the past even Newton accepted the religious view that time is absolute or fixed for the whole universe.

been lived through successive events since the 'Dawn of Existence', or intelligence, or the first acts of sentient beings on earth, from which acts all successive events have flowed as inevitable consequences, thus giving us the continuing story of human life on this planet. There was no 'dawn of time'; there's only the 'dawn of existence'. Today, time is not seen as marching on and taking us with it; rather we think we are marching on event by event and recording them as they occur at certain dates and times. These dates and times are the positions of the earth round the sun. A time unit is equivalent to a physical distance round the sun.

WHAT IS SECULAR TIME?

First of all, (before we discuss time after Einstein proper), we have to define what is meant by secular time despite this book's title. The Einstein theory of time is called 'secular' in the sense that it is traced or deduced consistently from premises based on material reality without any attribution to any gods or deities, and that is this: everything we refer to as time must be a recognised unit of time. The word time has no meaning without quantification---see Appendix 1, 'Time and Quantified Time' below. People usually mention the word time to mean the passage of anything---events, moments, even sitting still---but that is not logical thought or attempts to define time as it is used in all activities in society. Time in science, logic and life (or what of it that we need to understand or explain in ordinary linguistic usage), is presumed to be either materially based or mathematically calculated from features of the earth logically, as opposed to time that is simply imagined or mentioned as the term for any passing moments. One common example is time in dreams; another is just referring to time in ordinary conversations that are generally understood to mean any moment of passing consciousness---the shortest time, the longest time, and so forth.

In society or real life (as opposed to these vague instances of mentioning time without definition), it is necessary that everything we refer to as time---every unit of time in use---is logically traceable and derived from the periods of the revolutions of the earth and its long orbits of the sun---usually shortened to the phrase 'motions of the earth'. Not long ago the individual units of time, the hours and minutes and seconds, were regarded as divine; God had actually created them as independent entities.[154] They're easily explained with mathematics as fractions of the year that would not exist without it, yet the mischief-makers claimed they had independent existence. Even still now many scientists continue to speak of time units as if they're mysterious. But of course they're not. The year is determinate. All the units of time are also formulated as fractions of the year and its astronomical features so that a certain number of each unit will add up to exactly one year to coincide with the complete orbit of the earth round the sun. Thus there are no thirteen months; no 54 weeks; no 368 days in one year. The units do not go on after the year; they are all recounted from the base of one at the end of the year.

The 24-hour periods and the long orbit of the sun provide considerable natural duration for planning time for all activities: time to catch a Bus, plane or train; time to eat; time to walk a distance; time for work; time for sports---time for doing anything at all. All of these are derived from

---

[154] I concede that coming fresh to the study of time, the phrase 'time units' would require an explanation. But there is one available.

either the 24-hour periods or the earth-year. In secular time we realised that a mathematical or logical explanation was required, and our own Bertrand Russell, as the world's most recent great philosopher (who was also a logician, writer of genius and a great mathematician), provided the world with an appropriate theory called 'relation between points'; that's the only time that can be programmed into the clock, and we all know that time in the clock is the only reliable time.

Time, he said, is a construction.[155] Together with his collaborator, Professor A.N. Whitehead, he also interpreted the world of sense as 'a construction' rather than an inference, to overcome the old practice of philosophers inferring all things and connections in their minds as their logical definitions of physical reality, contrary to the physical reality discovered in physics, or the actual physical analysis of what we perceive and can also infer from what is perceived. So secular time refers to the time system we can consistently trace from the mathematical, logical and visual premises all combined. Every mention of the Einstein theory of time is to be understood as 'secular time'---traceable from material reality without mythologies. The only system of time we can programme into a clock. It is conceivable that Russell gained his insight by asking the question, if cosmic time is abandoned then what is put in the clock as time? Given sufficient logical acumen, everybody can deduce that the new theory is calculated (and can only be calculated) from the motions

---

[155] As previously stated. Leibniz also called time 'a succession', missing the term 'units'---time units in succession or procession as the answer to the problem of the passage of time. Nevertheless I mention this to show that Leibniz was also very clever, no wonder he and Newton quarrelled about mathematics; intellectually both were almost on the same level. We're lucky that these men come in doubles: Russell and Einstein, Newton and Leibniz, Plato and Aristotle, etc.

and physical features of the planet we live on. That's the meaning of secular time. Yet time belongs to human beings and not the cosmos; it is not part of cosmic reality, as I have already explained that what happens on planets (as implied in the concept and postulates of special relativity) cannot be relevant in the universe at large---and they include time as a conceptual phenomenon used for the control of events that are not recognised in the cosmos for several reasons. A natural law on earth is not in the same class as natural laws in the cosmos---the latter involve only massive objects, moons, planets, stars etc. Activities on the surfaces of planets, on the other hand, are so many billions and too small in size to matter in the cosmos.

TIME AFTER EINSTEIN

Albert Einstein changed the debate about time for good with his division of the universe into two distinct categories, governed by different natural laws: (1) General existence, or general relativity, where objects or matter just existed and whirled around under the influence of gravity without any conscious directions or time and order; and (2) special existence 'in' special relativity frames or bodies, where the two postulates and time applied, perspectives arise and intelligence and life can flourish in response to the intelligent use of available resources for civilizations to rise and fall---or generally for life to flourish as it cannot do in the whirling flux of general relativity; thus creating the never-ending chain of events known as history or the continuing story of human existence. Since civilizations arise upon definitions of time as mentioned above, the reader can see that only the very rational, scientific civilization can be consistent with the new concept of time.

Einstein did not deliberately set out to change our view of time. It was an accident discovered by Lorentz. He said the Lorentz concept of local time may be regarded as 'time, pure and simple'. His genius made it sound simple, but it was the beginning of the most profound revolution in human thought. And let me stress again that it is trough his insight that the solution of the passage of time occurred to me, namely if time is in units and units only the units are in procession to make it seem so smooth; and it also means its absorption is in units; since we live by time it means we absorb its units to live; they are instantly lost or gone on contact. Ergo, **the knowing of time is its usage and also the essence of its reality and passage through nature. So what is time but existence plus activity?**

Thus Einstein's notion of time was unique; the nearest idea pointing to the origin and purpose of life because time is the second most important thing in the universe, bar the life itself. And the two are inseparable. No thinker has any idea as to which is which or which of them came first. Personally I believe everything in human experience is generated by the brain in us; this implies that all human creations are secondary to the existence of life part of which is the brain.

Of course, on the other hand, Einstein did deliberately (and even contrary to classical physics), set out to change our views of the universe. The result was the theory of frames with which we are now familiar. It divided the universe into two distinct categories. As we all know, one is general relativity, where there is nowhere anywhere for life to evolve and flourish; the other is the inertial frame, where life is possible and civilizations can rise and fall. Time is required in this second division of the universe; and the local time idea was just the thing to suit inertial frames. I think we should now write time as the third postulate to add to the two original postulates of the special theory of relativity.

The Origin of Secular Time

The problem thence is to discover how our own time began, not as a version of a universal time, but a time limited to our frame; a major kind of philosophical inquiry since time is inseparable from life. That old idea was a mistake; yet everybody in science is still considering time as if it is something generally in existence and our time is a version of it.[156] Thus, the Minkowski formula for 4-D geometry is defined as incorporating time in the three dimensions of space to create 'space-time', the merging of space with time, the end result of which is to give us what we call time as 'space-time'! In the absence of a universal time, where is the time incorporated into the natural dimensions of space coming from, if the end result is only to create time again? Using time that does not exist naturally or universally to create space-time as time by means of mathematics--what sort of logical reasoning is that?[157] I am silly enough to let it bother me a lot; I really do not believe that it worries anybody else since everything I write is never even read in manuscript let alone published---my long suffering son has had to do it on my behalf, yet he's only an engineer! It may well be that people are literally afraid of time--- afraid to offend the Gods. Yet, actually, the nature of our time is easily deduced from elementary logic, and A.N. Whitehead, Albert Einstein, Bertrand Russell and Professor Eddington deserve the highest honours for the discovery. With all their boastfulness, none of the German Idealists ever even understood it all.

In fact, as Russell put it, "There is no longer a universal time..." Thus, he asked, "What is measured by a clock?" Yet the question is wrong. The

---

[156] Culturally it does not seem likely that we can ever abandon all traditional references to time; but in analysis scholars should try to do better.

[157] Yet this space-time is compounded of the materials found on the earth only for the convenience of the mathematician, according to Bertrand Russell, who, of course, did know a few things about mathematics.

clock does not measure time. It rather reproduces units of time specifically programmed into it for reproduction. That is the reason it works in units only---second, second, second, and so forth. The real problem is how the units of time programmed into the clock are derived in the absence of a universal time.

The year, of course, is basic. The seconds and all other units of time are derived from the year with mathematics as fractions thereof. That is why we calculate that 31, 536,000 seconds equal to one complete orbit of the sun or one year. Everything depends on the use of points. We use points to get the year. The fact that it is repeated over and over again to give us all the centuries means our time is determinate---in other words, our time is discrete. The essence of a discrete time is ended when the units are expended, thus we have to repeat the yearly cycle for our time to continue all the way to the centuries by replication. In addition, the system runs all through our units of time: from seconds to the minute, minutes to hours, hours to days and so forth. Our time is not a thread running through nature as we used to think; what we have found through experiments is that it consists of a chain of individual units created with points or mathematics in association with astronomy and the essential features of the globe; as such it consists of separate moments, as Professor A.N. Whitehead has confirmed.

It also means time is not known ahead; what we call time, say the year, is known after it has passed---e.g. the year is not ended until 31st December. That is when we can have a whole year. Then we have to start another year. The same principle applies to all the other units of time derived from the year, including, as I keep reminding the reader, the atomic units of time, because they have always to be related to the second to make sense.

The Origin of Secular Time

Secondly, time cannot be seen as the cause of events; events are physically caused; the times are recorded as the periods during which they occurred. Thirdly, time created with points and which is not part of a universal time, cannot have anything even remotely to do with what the religious leaders dream up about the nature of time. Fourthly, the passage of a time system produced with points unit by unit, as the year shows, requires no arrow or arrows to pass through nature: the units replicate to pass by---precisely as the years replicate to become centuries. All that remains for time to take its rightful place in science as a rational subject is for mankind to wean itself from the 'sweet' religious suggestions about time (what Professor Eddington called 'even-flowing time'), since the true facts are now well known: we don't know what it is, except to guess that it is the product of sentience, physio/chemistry and motion; but we know how it begins; we also know how it passes by---second, second, second; or year after year after year; and we know how it will end, that is, when our planet ceases to support sentient beings who can count the orbits of the sun as 'years'. Religion has nothing to do with it. The arrows of time for its passage through nature is redundant; and is definitely not universally existing in the cosmos because without knowing how to count the orbits of the sun as years, there could be no years just bland existence.

In conclusion, let me point out that, if the distortion from the Minkowski formula is eradicated, the question of time under relativity becomes simple, exactly as Einstein put it, namely 'pure and simple'. Here are the basic facts: (1) There is no longer a universal time so we have to search for the origins of our time because; (2) every 'body' or inertial frame has got to have its own time ; (3) under relativity the all embracing time is a construction, like the all-embracing space; (4) both the earth-year's time and the atomic time use regular or repetitive motions to track time----

that means they can only track passing time since the pulses or motions can be counted 'after' they have occurred and not before. We put all this together and get the notion that time cannot be logically defined, which means that what we call time are units of passing time. They are units because we get them from repetitive motions or cycles---and that is the reason it is passing time, simply because, of course, these regular cycles are passing. One after another (or year after another year), there is nothing more to time.

Thus, in the end, since the years are our only means of noting the passage of time, the explanation of time was rather easier than going through all those complicated mathematical and physical theories of arrows, mysticism and divinity. It is conceded that time is mysterious. It is even assumed to be the last refuge of God after Charles Darwin, since many people believe that time's deep, fearful and intimidating mystery goes beyond human comprehension.[158] Yet it is rather ironic that we have been using the orbits of the sun for time without realising that it is all of our time---mere physical cycles counted as years, centuries, millennia---because we thought we were measuring our version of time out of general time permeating the cosmos, the provenance of which was assumed to be nothing but divine. Yet once we learn that there is no longer a universal time (thanks to Einstein), and that we do not measure time at all, the orbits of the sun appear in a new light: we count the

---

[158] Time is the closest thing to the nature of life. Whoever solves the mystery of time must know more about life than the rest of us, and that honour goes to Albert Einstein for observing that the Lorentz 'local time' notion can be defined as time, 'pure and simple'. Until then everybody assumed that time originated from the cosmos and probably of divine provenance. However, if anybody can invent time, or his time, then its origins must be human. And that was a great philosophic insight.

mere physical orbits as our ultimate units of time (the years), out of which all other units are derived.

PHILOSOPHER/SCIENTISTS

Plato allowed for a creator. His theory of Ideas is one justification for a creator. But he was wrong. A Theory does not create reality; it can only reflect it through physical evidence. In the absence of such evidence, all theories are matters of opinion. Every suggestion from a human being without physical proof should never be accepted as worthy of attention. Plato has had adulation from religious people for far too long. Even Einstein never got a fraction of that kind of intellectual adulation; yet he rather had the necessary physical proofs. Human beings have been governed by some people's mad dreams for too long---chief among whom is Plato. In Hellenic times, as opposed to Plato, Lucretius rather was right. The quantum theory has proved him right: particles of matter are continually on the move even within a single atom. As they do so (or 'swerve') they accidentally cause chemistry; one result is life. Those academic philosophers still writing footnotes to Plato should be ashamed of themselves as Sir Karl Popper has observed.[159] In what follows, I have quoted passages from scientific writers and others from philosophers and have contrasted them to show how some philosophers

---

[159] What Sir Karl Popper is moaning about is nothing but academic philosophy, the very subject he taught his students throughout his illustrious career. What I think he is hoping for, due to the influence of Russell and Einstein, is an end to the intellectual habit of bypassing science. But if we are henceforth to give due respect to science, as Russell said, "on pain of death", then everything is bound to be footnotes to Einstein, particularly because of QED which is also (in theory) due to his paper *"On a Heuristic Point of View about the Creation and Conversion of Light."* Under QED light is the new reality, what logicians call WHAT IS.

(who are not followers of Bertrand Russell) continue to indulge in the fun of bypassing scientific activity as if it does not exist.

In the past everybody thought philosopher/scientists were born not made, but since relativity, the quantum theory and the electronic revolution, the internet, the computer and all the rest of it, every aspects of life has become so complex that we need to train our own philosopher/scientists to guide us through the maze. Socially too things are getting out of hand, so we have to be careful or life can be extinguished very easily through the actions of a few mad men. The artificial, classroom philosopher/scientists we can create may not be as brilliant as the originals, certainly not, but equally we cannot go on like this without guidance---for instance, relativity is still no properly understood. Mathematicians claim that it is only with the Minkowski formula they could make sense of it; on the other hand logicians insists that since the Minkowski formula relies on 'i' or imaginary time, it cannot be used to determine the nature of physical reality as to whether or not the physical world or physical reality is four-dimensional. So what is happening now is a scientific anomaly because all science is based on the concept of four-dimensional space, as space-time, yet logically the notion is fatally flawed. I count myself among the very few people who are writing against it amid mockery.

Already we are able to train scientists and philosophers; even self-educated people have to learn their ideas from books; that is also training of sorts, except that self-educated people have to work harder. All educationists know that what a professor can easily impart to his students in one lecture would take the self-educated person ages to glean from books. Nevertheless, we train all our experts (physicists, astronomers, mathematicians, biologists, bio-chemists, architects, engineers and dozens of others in technically demanding professions.) They don't come from above; training makes them what they are or what they turn out to be. Many of them go on to teach others in schools, colleges and universities. It is true we don't know how to train people to make discoveries and inventions or propound profound theories about the universe and the world we live in. In a word, we do not know how to create geniuses. Still we do from time to time get some people to whom discovering something out of the usual or from thin air involves sacrifices they would readily subject themselves and their families to without thinking of the money, fame or even the preservation of their lives. From such persons we get our ideas, useful ideas, I must stress, and so I will quote three passages about scientific facts and other three about philosophical ideas imparting great wisdom to illustrate what knowledge means to us. This is a book about time based on relativity, so I think it is right to demonstrate the merits of the scientific outlook as against the Platonic mysticism.

First, a scientific piece (not a mythological one in ancient man's style) about the sun itself as the source of life on earth: "Most of the energy of the sun comes to us in the form of light. Sunlight is transmission of energy, in the form of electromagnetic radiations from the sun to the earth. When Stephenson's first crude steam locomotive was moving along its wooden track, the inventor asked one of his companions what was driving it. 'Your engineer from Newcastle, I would say.' 'Wrong', replied Stephenson, 'the sun is driving it.' I suspect that many millions of us who race the roads in our automobiles have not yet grasped the meaning of this simple sentence. Most of the mechanical work of the modern industrial world is done by energy stored in fossil fuels. Other power comes from water lifted aloft as it flows downwards to the sea...Visible light is the most familiar form of the radiant energy that reaches us from the vast and distant sphere whose surface temperature is estimated to be about 10,000F..."[160] These statements are all facts, even though they may sound strange to the layman. But how have they been pinned together? The answer is bit by bit over several centuries and by numerous contributors. The important thing is that they are true or cannot be refuted without further research. The research will have to be scientific; you could not do it sitting comfortably in your armchair. But equally you do not need to go into a laboratory to do the necessary research. It is a question of attitude. Your thinking must be scientific as stated by Bertrand Russell in his essay "The Rise of Science", namely: "It is not what the man of science believes that distinguishes him, but how and why he believes it..."[161] How he believes it is more important---how he believes what he does, what actions did he take to lead him to those concepts? It is reported in the Press that many people were surprised by the success of the recent science-based stage play Night of 200 Billion Stars, but I wasn't. Rather I was delighted.

After a hundred years of Einstein and several years of encouragement from Bertrand Russell about "The Scientific Outlook", followed by the computer and internet, mobile phones and all, it would rather be surprising if somebody did not write a science-based play of the sort and get acclaimed for it. For science is becoming popular, particularly through space travel, as many mysteries of the universe (or a few of them!), and also wonders of the sub-atomic level are revealed, especially with the magical qualities of the quantum (or a few of them) thrown in. Many books have even been trying to claim that, because of the Minkowski equation of space with time from which they assume the

---

[160] Professor Paul B Sears, *The Biology of the Living Landscape*, Allen & Unwin, 1962.
[161] History of Western Philosophy, Book Three, Ch. 6.

existence of something called "curved space time", time travel is a "scientific possibility"; and although they are wrong because the Minkowski theory is false, the layman is bound to be intrigued.

It's not very clever describing science as just one of the many equally valid ways of looking at the world without scrutinising why, how and what the scientist believes or examines the world for. His job is the logical and systematic search of what is really there in nature and which can be employed to serve the life of man to make it comfortable, longer and happy; for although death awaits us all, scientists want to make human life longer, safer and more comfortable and happy--- by and large, the evils of science come from politicians (some of whom are the most devious of human beings), not the scientific researchers.[162] And one thing that the religions and the anti-science brigades forget is that science is progressive; we wrestle scientific or dependable knowledge form barren, even hostile, nature. Take scientific medicine as a prime example. Obviously it evolved. Man did not know a thing about scientific medicine when life began. It only gradually and even painfully evolved from the researches (and the research methods had to be learnt) of countless individuals, many of whom never lived to enjoy the fruits of their labour. So now that we have grown to know scientific medicine, go to the moon and beyond in search of more scientific knowledge to use for the improvement of life on earth (for instance, astronomy may seem to be a mere academic exercise, but it is from astronomy we are learning how to deflect or explode asteroids likely to end life on earth),

---

[162] Of course, a few individual scientists may be devilish, but, on the whole, science is for the benefit of mankind not its destruction and even individual treachery is often perpetrated in extreme secrecy simply because the scientific community could destroy the traitor without mercy on the general understanding that science is for salvation.

we can now eradicate many diseases including polio and small pox;[163] we have also invented the computer, the internet and wireless communication. So, I repeat, it is not very clever to say science is but only one of many equally valid ways of looking at the world---it may be so, but only systematic, logical, scientific knowledge brings lasting benefits. In that sphere there is no rival to science. Besides, nobody knows what is 'the valid' way of looking at the world. Even scientists make no such claims. Their method is to use our most incisive organ (the eye) to observe and report what they see and use them for human salvation. They are not claiming to be capable of doing more than this--- but it is enough. It gets us most of the things we need for normal life.

Next, I quote from the popular book, Human Situation, by Professor Macneile Dixon[164]: "In the great arch of night above our heads about five thousand stars may be seen by the naked eye. In their marchings and counter-marchings they make a brave show, yet are in fact scarcely so much as a swarm of bees in all Asia, a spray of blossom in the limitless abyss, where 'a hundred thousand million stars make one galaxy, and a hundred thousand million galaxies the universe'. The stars we see are but a handful, and their removal would not disturb by as much as a decimal the calculations of the angle of their courses. We may be sure that for every human being in the world there is not one star apiece--- there are ten thousand. Viewed from the bodily angle, no comparisons can express the insignificance of man among the cosmic magnitudes upon which our astronomers exhaust their eloquence...The earth is a mote of dust, and the sun itself a diminutive firefly. We inhabit the puny

---

[163] I say 'we can' do so; more optimistic experts claim that we have already achieved the total eradication of these terrible diseases---but I am still keeping my fingers cross that some terrorists will not be able to resurrect them.
[164] In its day this book was almost as popular as the Bible.

satellite of an inferior orb. There are millions of stars so immense that room could be found for millions of our petty sun in one of them."[165] This was written nearly a hundred years ago. We have to update the figures. According to the Philip's Concise World Atlas, p.3, "At least a billion galaxies are scattered through the Universe, though the discoveries made by the Hubble Space Telescope suggest that there may be far more than once thought, and some estimates are as high as 100 billion. The largest galaxies contain trillions of stars, while small ones contain less than a billion." If we can train researchers to establish such complex and yet accurate details of the universe, then we can also train them to do the other parts of philosophers' work, for this is also philosophy.

Scientists or astronomers specifically, can now speak with greater authority about the universe than philosophers. To put it another way, what used to be the exclusive preserve of technical philosophy, that of telling us the nature and composition of the universe, as a matter of speculation or inference, are now displayed openly in elementary books by astronomers with proofs or all the necessary physical evidence required.

For our third example of scientific truths, as opposed to philosophical speculations, on the understanding that man requires both for his material and intellectual advancement, I quote from the New Scientist: "QUANTUM electrodynamics is arguably the most successful scientific theory there has ever been. With stunning precision, it explains the interaction of electromagnetic radiation (including light) with electrons and other charged particles. It is on QED that quantum

---

[165] *The Human Situation*, Edward Arnold, London, 1937, p. 155.

chromodynamics, the theory of the strong interaction, is modelled."[166] This is the solid ground upon which Quantum mechanics is built. Two Nobel Prize winners put the same idea in different words. In case the Magazine presentation strikes any readers as down-market, I will presently quote them for their satisfaction. First, Professor Richard Feynman began chapter 4 of his book, QED, saying: "...I am going to talk about problems associated with the theory of quantum electrodynamics itself, supposing that all there is in the world is electrons and photons..." and the wonderful Louis de Broglie also said: "Without the Quanta was not anything made that was made." Ordinarily we know light as immaterial. These statements are not only shocking but border on the ridiculous. Yet they are true, and philosophers who argue against them rather make themselves ridiculous. The religions may object because they deliberately like to object to scientific progress as it undermines their beliefs and make them the laughing stock of modern man---and they will tell you that the more they are mocked the more they like it because it tells them that they are having some effects, whereas they know that religious talk should not be heard at all by people dealing with true knowledge. But a philosopher, as a man of learning, the lover of wisdom, cannot contradict science with mere assertions derived from his arm-chair conclusions. He must incorporate the discoveries of science in his thoughts. There is no way he could do that unless he learns to become a Philosopher/Scientist, which refers to somebody who reasons by means of scientific facts only. So let's go on to show examples of philosophical sayings that fail this test.

First, I quote from the Oxford philosopher Professor William Kneal's book On Having Mind (Cambridge, 1962.) In concluding a small book

---

[166] From The *New Scientist*, 8th Jan. 1994.

about how and why we have minds, he wrote: "We must retain the Platonic notion of mental events which are distinct from anything in the physical world and manifest a special kind of connectedness." So, according to this 'learned' professor, there is a non-physical world in addition to the physical one we live in, and because of that the Platonic notion of mental events (which deals with that mythical world), is the true theory of how and why we have minds.[167]

In fact, the quantum theory undermines the Platonic idea. The invisible world is that of the quantum, namely images of things can be cut if the lights from any object do not reach the eye. Bishop Berkeley has already proved this without even knowing it, and I will come to that in a moment. But his 'proof' allows us to infer that light radiation from objects conveys the images as the surface silhouettes of objects for us to see them, for we know that the particles of light, the photons, are naturally coloured. In an essay on Bishop Berkeley, in his monumental book History of western Philosophy (which I urge the reader to consult about this matter), Bertrand Russell wrote: "Berkeley advances valid arguments in favour of a certain important conclusion, though not quite in favour of the conclusion that he thinks he is proving. He thinks he is

---

[167] By mental events he is not telling us about the minute, hidden and invisible causes which are also physical, that prop-up bulky matter. He is rather bizarrely referring to something akin to magic in homage to Plato instead of observation. He is dreaming rather than looking at the world to see what is there. And he's so sure of himself that he insists that 'we must' do what he prescribes! With professors like this in our great institutions writing this kind of nonsense about philosophy, who can blame scientists for laughing at philosophers? Either we call what is done in philosophy departments 'logic', and restricted to logical studies or philosophers should pay attention to science rather than Plato's seductive mysticism.

proving that all reality is mental; what he is proving is that we perceive qualities, not things, and that qualities are relative to the percipient."

My next example of how philosophers say things about reality to contradict (and therefore reject) what scientists actually find 'out there' is taken from a review of the book, "Science, Perception and Reality", by Professor Wilfred Sellars, an august Harvard professor of philosophy, a man of learning or of wisdom. Only it turns out that what he knows is scientifically rubbish and not worth a farthing:"Professor Sellars nowhere states the purpose of his book, and, since it is a collection of independent essays on a variety of topics, its intention can be judged only from its title. It is therefore not unfair to take it as an attempt at a philosophy of science...From this point of view, its value is slight. It reverts wholeheartedly to the Mill type of bypassing scientific activity, and analyses questions which are quite independent of anything scientists do. To take but one example: 'Philosophers have been fascinated by the fact that one can't have the concept of white without being able to see things as white, indeed, until one has actually seen something white. But this can be explained without assuming that sensation is a consciousness, for example, of white things as white". The reviewer, Professor Herbert Dingle comments, "A Scientist could scarcely care less for what has fascinated philosophers. He does not regard this as something to be explained. He starts with observation and forms concepts as required to express the relations he finds between them..."[168]

My third quotation is the withering criticism of philosophers by another philosopher (mentioned before), only this one is a follower of Bertrand

---

[168] Professor Herbert Dingle, reviewing *Science, Perception and Reality*, TLS 25th October, 1963.

Russell, Sir Karl Popper: "You see, the history of man is a queer thing. It's a history of a succession of attacks of intellectual madness, of all sorts of strange intellectual fashions. I don't need to give many examples of revolts against reason (such as Existentialism), for we know how strongly certain fashions have taken hold, not only just in a comparatively small insular group, but, in large parts of mankind. Russell saw these things in that light, and so do I...In the long history of philosophy there are many more philosophical arguments of which I feel ashamed than philosophical arguments of which I am proud...Yes, I cannot say I am proud of being called a philosopher..."[169]

I have quoted extracts from the works of scientists and philosophers. I believe it will not be difficult to decide which of these thinkers are speaking the truth, especially given the harsh criticism of philosophers by Sir Karl Popper, one of the great thinkers of the 20th century. Traditionally we are told that the purpose of philosophy is to think about the universe and the world we live in. If that thinking exercise is so bad that one of the leaders in the field is ashamed to be called a philosopher, then we have to agree that something is wrong with either philosophy or how we train our philosophers. I think both are misguided: The traditional topics discussed by philosophers are no longer relevant to the world we live in, and how we train them is also archaic. Of course, it is common knowledge that since Einstein and Bertrand Russell some institutions have started to call what they do "The Philosophy of Science." But I am not convinced because when I browse through some

---

[169] Sir Karl Popper, in conversation with Strawson, Warnock and Bryan Magee, published in *Modern British Philosophy*, London, Secker & Warburg, 1971. Sir Karl Popper was sending out a message because he knew that what he was saying would be published in a book---I've mentioned this above, but this is the full text.

philosophical journals I see that they are publishing articles on the same old traditional topics from Plato to Kant, with particular emphasis on the history of the subject without stressing what is right and what is wrong in the ideas of previous philosophers. Yet that is what made Bertrand Russell's History of Western Philosophy particularly valuable. The old philosophies remain footnotes to Plato, by whom there are two worlds: the visible world of physics and the invisible realm science cannot reach, as Professor William Kneal put it; it consists of the world of "mental events which are distinct from anything in the physical world..." So mental events are to be preferred to actual physical events occurring out there and influencing our lives physically.[170] This is anathema to scientists who are already baffled by the quantum theory, which amounts to the analysis of matter down and down to the invisible sub-atomic particles. In fact it does appear that material physics is finished; the physical analysis of matter is at an end.[171] There is no longer any solid matter to analyse as it has been realised that solid matter is composed of invisible matter down to nothingness.

Of course there is a problem with induction, as there are numerous problems with all aspects of life, especially in medicine. What scientists are saying is that they don't know everything; so many of the quandaries in life defy scientific explanation. But they don't allow them to frustrate

---

[170] What scientists are asking for is for the resources to go out there, see what is there and study it for solving human problems, as against philosophers telling them to honour the Platonic mysticism.

[171] There is no doubt, of course, that technological innovations will never end. The things that can be invented out of the particles known to exist are literally infinite. At present we need inventors more than mathematical theories in all branches of science; but the idea that Plato's theory can be regarded as the precursor of the sib-atomic world is nonsense. The sub-atomic world is not mental; it is physically existing out there.

scientific activity. They rather get on with it, and they are right because they get results. As Bertrand Russell has warned, in physics for instance, we have to obey scientists on pain of death, because whether traditional philosophy approves or not the scientific method can destroy life.

On the contrary, instead of doing everything in accordance with what the traditional philosophers have said or are still telling us to do[172], Russell was bold and commendably judgemental; he condemned some thinkers; he praised others, almost precisely as Sir Karl Popper has also done, because Popper was one of the 'great' followers of Russell. It must be stressed that the notion that philosophy has to move closer to science has been known for years. Let me illustrate what I mean with a few quotations from Russell's great book. I remind the reader that we are discussing how Russell, guided by logic and history, science and common sense, praised or condemned some philosophers, and I regard that as one of the best things he did. It is extremely important to judge philosophical ideas by the requirements of human welfare and survival. For it is not disputed that philosophy is necessary; the contention is that, as the quotation from Professor Kneal has shown, not many of them respect scientific ideas or the scientific way of looking at the world, ordinary human welfare and what is necessary for human survival.

About John Mill Russell wrote: "John Stuart Mill, in his Utilitarianism, offers an argument which is so fallacious that it is hard to understand how he can have thought it valid."[173] On Aristotle he wrote: "In reading

---

[172] This is the trouble with some religions. They do not want to move ahead but rather repeat what their ancient books are telling them to do, however ridiculous in our present modern world.

[173] Poor old Mill. One of the lessons I have learnt is that the modern view that philosophers merely debate topics as intellectual exercise using logic is not correct. I seem to sense that all philosophers want to be remembered for

any important philosopher, but most of all in reading Aristotle, it is necessary to study him in two ways: with reference to his predecessors, and with reference to his successors. In the former aspect, Aristotle's merits are enormous. In the latter, his demerits are equally enormous. For his demerits, however, his successors are more responsible than he is." Thus, in spite of his many faults, especially in his Metaphysics, Aristotle got off lightly. But on Rousseau Russell was clear that he invented evil: "He is the father of the romantic movement. Hitler is an outcome of Rousseau, Roosevelt and Churchill, of Locke." To Russell, Nietzsche was also a merchant of evil, and who can blame him? He said: "I dislike Nietzsche because he likes the contemplation of pain, because he erects conceit into duty, because the men whom he most admires are conquerors, whose glory is cleverness in causing men to die." But Hobbes and Bacon came in for some praise. He wrote, "Hobbes (1588-1679) is a philosopher whom it is difficult to classify. He was an empiricist, like Locke, Berkeley, and Hume, but unlike them he was an admirer of mathematical method, not only pure mathematics, but in its applications." "Francis Bacon (1561-1626), although his philosophy is in many ways unsatisfactory, has permanent importance as the founder of modern inductive method and the pioneer in the attempt at logical systematization of scientific procedure." Also he said: "Spinoza (1632-77) is the noblest and most lovable of the great philosophers." On Schopenhauer he wrote, "Historically, two things are important about Schopenhauer: his pessimism and his doctrine that will is superior to knowledge. His pessimism made it possible for men to take to

---

discovering or solving something, saying something intelligent. Thus why some of them ignore science is difficult to understand. For scientists also set out to solve problems, using observation so that what they find can be repeated by others. If we train scholars to combine the two disciplines collectively man will come to command a powerful mind.

philosophy without having to persuade themselves that all evil can be explained away, and in this way, as an antidote, it was useful. From a scientific point of view, optimism and pessimism are alike objectionable: optimism assumes, or attempts to prove, that the universe exists to please us, and pessimism, that it exists to displease us. Scientifically, there is no evidence that it is concerned with us one way or the other." Lastly, let me conclude this section with his views about Rene Descartes, who else?[174] The man who made French intellectually as great as those of ancient Greece! Of course he is praised by Russell, and rightly so: "Rene Descartes (1596-1650) is usually considered the founder of modern philosophy, and, I think, rightly. He is the first man of high philosophic capacity whose outlook is profoundly affected by the new physics and astronomy. While it is true that he retains much of scholasticism, he does not accept foundations laid by predecessors, but endeavours to construct a complete philosophic edifice de novo. This had not happened since Aristotle, and is a sign of the new self-confidence that resulted from the progress of science." This is a description of a Philosopher/Scientist, not simply a thinker in the mould of Professor William Kneal and his 'mental events' or Professor Wilfred Sellars, who, according to Herbert Dingle, "...reverts wholeheartedly to the Mill type of bypassing of scientific activity."

---

[174] Einstein was not the only philosopher/scientist. Aristotle, Descartes, Locke, Russell, Sir James Jeans, Professor Eddington, Professor A.N. Whitehead among others, were all philosopher/scientists. In my judgement Einstein was the most famous, Aristotle perhaps the greatest and Russell was the most popular. I believe it all goes back to Aristotle. After he taught us about the "logic of things", the cleverest minds realised that those studying the logic of things and those theorising about things in metaphysics ought to have ways of complementing each other's work.

Science is always good unless it is deliberately misused by some evil men, mostly for political purposes. Otherwise the basic aim of science is human salvation. However, as we have seen, philosophy is not always good, but very, very important because scientists cannot adequately think about the value of what they do; yet there are thousands of them; theories abound; some contradict others. It is necessary that a class of clever persons, or thinkers (suitably trained), makes it as part of its business to look at what scientists do overall and advise them in the human interest. I believe only scholars who philosophise can do that. They do not necessarily have to have doctorates from august universities, but they must think as philosophers not so much through speculations as through what scientists are finding out about people, the world and the universe at large.

Training, or education, makes a man. Lawyers coming out of law schools, as a prime example, always seem to have been brainwashed to be instinctively against crime, democratic and fair in their judgements about human frailty and dead against torture, unless they are basically evil in nature. Otherwise even when a person is found guilty of a heinous crime, his lawyer would plead mitigation and mercy for him or her. They even campaign for improvements in prison conditions. Why can't we train philosopher/Scientists as well---that is, to be equally fanatical about science in its noble pursuit of reliable truths in the physical world, medicine and society itself? To put it another way, why can't we devise a system of training to make competent people, men and women, capable of understanding science and instinctively think about it philosophically. Let me suggest how we could go about this training, on the understanding that I am a fallible human being and that my suggestions may not be the best imaginable. However, somebody has to start the ball rolling.

The Origin of Secular Time

First, the philosophers: actually nobody can prescribe what must be taught to scholars to make them Philosopher/Scientists forever. The subjects will keep changing. But, on the whole, as we have seen, if philosophers concentrate on the Platonic mental events so that they can only write footnotes to Plato; or, conversely, if they chose the Mill type of bypassing scientific activity, they would have nothing of interest to say to scientists. Yet science dominates every aspect of human life and must be given competent intellectual, or philosophical, guidance. With this proviso, and roughly speaking, I think that for future philosophers to be able to understand science properly, they will have to be taught, among other things, subjects like logic and mathematics, metaphysics and astronomy, ethics and psychology, and also the general principles of scientific medicine. They cannot ignore history and literature, meaning the works of great writers and literary criticism.[175] All the sciences, particularly quantum physics, will have to be taught in philosophical classes. Bertrand Russell's books must be read, and read very well. Methods must be found for teaching Relativity, special and general. It must be taught without the Minkowski contribution; then, conversely, taught with it, so that scholarly will come to understand the 3+1 formula as opposed to the Minkowski purported equation of space with time to abolish the 3+1 system and contrast the differences. By showing students what is meant by merging space with time and why the

---

[175] C. P. Snow's 'Two Cultures' should become a textbook. Writers like Noam Chomsky and linguistic philosophy should also be taught. Science is not a hobby. Science is not for scientists alone. It is for everybody's benefit. It is necessary to communicate it properly to all and sundry. Professor Sir Arthur Eddington, Sir James Jeans, Karl Popper and others including A J Ayer, A. J. P. Taylor and Professor Macneile Dixon and his *Human Situation* must all be taught for their intellectual achievements, their humanity and their literary merits.

Minkowski technique for doing so is untenable because of his use of imaginary time coordinates, they will come to understand that reality is based on observation, not on somebody's mathematics alone.[176] For if physical reality is 3+1, mathematics cannot change it to one of four-dimensional continuum.[177] The above tentative suggestions may be found useful in planning the philosophical education of future generations of philosopher/scientists. But they must also learn about ethics, economics and politics, particularly about the dangers of 'Failed States', both external and civil wars, the importance of peace and constitutional law---these subjects show how important thinkers can be. Now let us look at the education of scientists to make them appreciate philosophy and philosophers.

The majority of scientists are woefully ignorant of the other branches of science. Also, it is they who are moaning that they cannot understand relativity without the Minkowski formula, which means that, since the Minkowski formula is logically flawed, they do not properly understand relativity. So the above training programme plus their own specialist fields will be required---not much else needs to be said on this subject. The only problem concerns the complex mathematical interpretations of general relativity based on the concept of "curved-space-time". By this monster time travel is said to be possible. Of course, if space is the same thing as time then it could be. But the fact is, if the Minkowski equation of space to time is false, then when space curves it cannot take time with it. The whole concept will have to be re-examined on the basis of

---

[176] Thus they will also come to understand what is the coordinate system in modern physics.

[177] It creates discrepancy between the mathematical symbols and the essence of physical reality---my main objection to the Minkowski formula for equating space to time.

the 3+1 formula because it means four-dimensional space or 4D geometry does not exist, yet scientists are propounding all their theories on the basis of four-dimensional space or space-time. It is a serious problem and I don't know how they are going to solve it, for so many theories since Einstein will have to be reviewed. In plain language, many scientists are going to have to try to understand relativity without the concept of 4-D geometry, or the idea that space and time constitute one entity---a lot of scientific theories are going to be discarded as happened after Einstein when the eather hypothesis was abandoned.

PART FOUR

THE STATUS OF EARTH TIME IN THE UNIVERSE

I mentioned this topic briefly above, now here are my arguments.[178] First, the definition of time so that we know what we're talking about: it is not a time system that runs through the whole universe, for there is no longer a universal time. Perhaps it will do no harm to quote Russell on this again: "There is no longer a universal time which can be applied without ambiguity to any part of the universe; there are only the various 'proper' times of the various bodies in the universe."[179] The only time we know of is earth time, which is based on the orbits of the sun and is strictly determinate. The year has to be repeated for our time to continue, otherwise there is only one year. That plainly means we have only discrete time of limited application, but the cosmos has no time at all, and what appears to be time in interstellar space is caused by either

---

[178] The heading is enough to start the reader thinking that our time cannot be applied to the rest of the cosmos as our noble Bertrand Russell stated in his ABC of Relativity, for the cosmos is simply too vast and complex; with so many hundreds of billions of mighty stars, our tiny lump of dirty rock cannot invent a time system applicable to the whole universe.

[179] Bertrand Russell, ABC of Relativity, Ch. 5. So when I castigate some cosmologists for claiming they had proved that time will end in a black hole, I have good grounds for doing so. Time is not running through the universe to end anywhere. Whatever mathematics are produced (and mathematics can be used to prove anything), there is no sense in saying time will end in a black hole when there is no universal time---meaning apart from the time in your clock, there is no time anywhere else. And the time in your clock does not march; like the years it replicates its units to pass by. It means that reality consists of successive images or moments of contacts.

chemistry or accidents, which we may generally refer to as the natural state of existence in the inanimate world, where brute force is king.

Here on earth the whole idea of time---as complicated and mysterious as it is---is simply a mechanism for showing 'how long it takes' or 'how long it lasts', but of what? It surely must be 'contact'. Again time is reality as perceived even as it is changing continuously at the atomic level, such that a digital camera can capture several images in an instant---all different at the atomic level. The actual images of reality are converted to 'units of time' by the repetitive cycles we use to reckon time. So while the time itself does not move, the movements of the cycles create the illusion of the march of time. Time's actual movement is achieved through replication of its units---the years, for instance, replicate to become the centuries; it is the same with all the units of time which are fractions of the year. The 'March of time' concept is extended to history, reckoned from the last or past minutes. Yet history is rather the march of events not time. As the mere apprehension of reality, time does not march at all. It is contact with reality as it is. Whether your contact (or what you perceive) will last a minute or hours depends on the cycles used for reckoning your time. We on earth all have the same units of time because we use one cycle (the year) to reckon our time.

Without contact there is no need to know anything because there will be no evidence of anything's existence other than oneself as a solitary 'Being' (with no need for time reckoning.) So, given that it is man against the rest of nature, contact implies everything external other than ourselves as individuals. Even then touching parts of one's body is also contact for the reckoning of time expended. This or these contacts begin from the womb; hence the sense of time is intrinsically part of our nature. This may be the reason time is inseparable from life and

232

extremely difficult to explain as 'constructed' by man rather than divinely bestowed from the heavens.[180] Yet scientific or logical thought has absolutely no truck with mysticism. Either we know something or we don't; there is no need to introduce revealed knowledge into discussions of material reality. Of course, I am aware that many scholars believe that mysticism has a role in scientific thought, but Bertrand Russell thought otherwise, and therefore did not mention Wittgenstein even once in his great book on Western philosophy, because his work amounts to nothing but logical mysticism which aims to put an end to physics. This is as close as calling him insane.

After the definition of time (without mysticism, revealed knowledge or fantasy), let us begin this section of the book with one important statement of fact about time: time consists of units---or single moments---as Professor Whitehead prefers to put it. The cardinal example is the earth-year. However it is utterly impossible to define any unit of time in the abstract such that it can be recognised when conditions for it are fulfilled. In other words, it is impossible to know what conditions will produce (exactly) what unit of time in any part of the universe. The year

---

[180] Quite frankly time is not only difficult but almost impossible to discuss as a logical subject. I have heard stories about murderous dictators doing away with thinkers who speculated about time, and I believe them. The reason is that the whole of human existence, culture, civilisation and even the very air we breathe, and for what purpose, is changed when time is seen as anything but traditional time, for otherwise the thinker has to explain why and how everything in the cosmos came to be what they are and when----also, for what purpose? On the other hand, ever since Einstein showed that absolute time cannot exist, the traditional concept of time has become logically untenable as one time system covering the whole universe so that a second here is the same everywhere. Thus there is a dilemma, and it makes writing about time the most thankless undertaking in the world. Nobody will even look at what I write. Nothing personal; rather they think they know what time is and have no problem with it.

itself can never be defined in logic. Therefore what a second of time is can never be defined in such a manner that people can recognise a second when they see it. For this reason alone earth time cannot be valid in any other frame, including that of general relativity, and I think cosmologists need to be reminded of that. Now let us consider the status of earth time in the universe, knowing that it cannot be applied anywhere else, being a product of the earth's peculiar postulates and periodicities. I have come to the conclusion that it is possible our time is created with a unique set of circumstances, and that on other planets it may not be the same or even called 'time'.

Now, how life appeared on this planet, and for what purpose if any, as a composite question, most dire, is the only conundrum greater than how we get the time by which man lives his life, and how the time moves on, or continues perpetually. It is not surprising that before the rise of science (and even still now) many thinkers believed that it is time that gives us the right to live;[181] the general idea was that life is based on time; whereas the scientific study of time indicates that time is based on life---somebody must be there to count the orbits of the sun as years (consisting of the seconds and all the rest of it), or there will be no years and no seconds, only bland existence in a senseless world. Time is the motive force in the mind. Every movement, thought, action and inaction

---

[181] Strictly speaking, it is true that one's body gives him or her the time to live in the sense that your physiological make-up determined how long you will live. In the study of time, however, it is the other way round: time is based on life, for it is the living who counts the external cycles as years, otherwise there will be no years. So the time system is based on life, but the life has a span that can be stated in the language of time. The effect is to feel as if one is given 'a certain amount of time to live' at the end of which death will 'lay its icy hand' on him or her. In fact, the span is strictly determined by one's physiology, and can be manipulated or managed, because it is not fixed.

occurs in time. Time enables us to act and be human since every action takes place in time.[182] So if time is created by man then it must be two in kind: the internal sense of time and the external parameters by which we measure intervals. The sense of duration (the internal sense of time) is natural; what we create to link it to physical reality is the external time, and is strictly based on points as applied to space. Is this man-made time any use in the cosmos at large? I think not, and will discuss it in a moment. This brings to the fore for a brief discussion the question of whether or not there is time (or there will be time) in the absence of somebody counting the orbits of the sun as years. As a question for serious debate it is akin to the problem of the existence of God, because it is so mysterious and of great philosophical significance. Is there time in the absence of human intelligence? I think not. Without the human mind what appears to be time is rather the natural duration of events caused by chemistry or accidents.

My personal view is this: there is always motion of the kind we associate with time, like things moving on, growth, decay and ageing, even leaves on trees waving in the air. They are also action and they take time. Some thinkers assume this process to be time moving on, perhaps even silently, without the intervention of human intelligence, or rather not even requiring man and his mind at all; that, naturally, trees will take time to grow, for instance. As a matter of fact, such natural events do not constitute time per se; instead, I see them as events occurring to certain objects in their own worlds, or in their own 'Beings': a tree grows, a river flows, a person is ageing, and so forth. It is not time we can mechanise in a clock. Things live their own lives to which time, once

---

[182] Obviously this is chemistry, action or any motion, all of which we call 'natural time', as opposed to constructive or quantified time---the time that can be used to build civilizations..

mechanised in a clock, can be applied; but the passing of their growth or decay is not quintessential time. It is not constructed or quantified time.

Although the general growth of a tree can be explained as "time going". It is not the kind of repetitive cycles we can mechanise in a clock. All things that move or grow can be set to mechanised time; whatever happens to them happens through the passage of time. But it is not correct to call the invisible growth of the human hair as time. It is not time but the natural chemistry of the human hair. Rather it is correct to regard it as "time going"---invisible time going. To know how much time in physical and visible reality you have got to rely on mechanised time based on regular or repetitive cycles and count them as the rates of time, the years, for instance. If we regard every growth and motion, backwards and forwards, as time we would end up virtually in a confused world of myriad of time systems. So for scientific and logical thought we rely on mechanised time; all references to time should be reserved for mechanised time suitable for universal application in one inertial frame. (See 'Time and Quantified Time' in Appendix I below.)

Again, if we consider any motion as 'time' rather than as 'time going', the implication will be a reference to pre-existing universal time, not one created locally for local purposes as space-time in specific units which can only advance 'unit-by-unit'. On the other hand, the idea of taking any motion as 'time going' (in units), means the time must have been established already so as to have it in specific units. All references to the concept of space-time are to be understood as time obtained from space with points, not in the sense that time and space constitute one entity by the Minkowski mathematics---the man who succeeded in fooling the rest of mankind with mathematics, He's not a fraud except that mathematicians can be gullible, even childlike. Another one got up to claim he'd made a discovery that time will end in a black hole, when the time had already been proved to be discrete, no longer universal and utterly impossible to travel through the universe to end up in a black hole unless it's carried by mathematicians on their bald heads.

It has happened in the past that many things were used to mark time: the shadows of trees, of mountains, of houses, even human shadows. But since the earth-year is now used for time over the whole planet (albeit with zonal variations), we tend to interpret motion and events in terms of the amount, and length, of units of time expended, being expended, or, futuristically, to be expended. It is the same under the relativity notion of time. Time is now known as 'space-time', being the product of points as applied to space; space-time is necessarily discrete; it does not run through all nature. As such we have to search for the method used to establish our time as space-time, which, of course, is (and has to be) limited in its effects to this planet alone. In this situation, the only way to have one time for any inertial frame overall is to mechanise some repetitive cycles (the year, for instance) as time for all and sundry---with the inevitable zonal variations, depending on the size of the frame or planet. The clear philosophical implication in epistemology is that time does not exist in nature at all, if time is defined as "time in the clock". Yet only time in the clock is scientifically relevant. Sentience is required; intelligence is necessary; the ability to count is indispensable; and a theory of numbers is absolutely essential, all of which makes it seem human in origin, but based on the natural sense of time as "duration" felt in the mind. It should be noted however that duration also requires points, therefore contact is still required or implied as the origin of the sense of duration and time.

The Origin of Secular Time

Even the concept known as "the passage of existence" is meaningful only as "passing through a human mind". Somebody must be there to count the orbits of the sun as years, and have the intelligence to sub-divide the year down to the seconds. Thus, precisely like language, time arose out of necessity. The multifarious activities, motions, and events in existence (growth, decay, to and from, up and down, backwards and forwards, and so forth) all occur to individual things and beings---human and animal, rock and plant. They are individual occurrences; things living out their own lives through their physiology and chemistry, physical and organic. Otherwise there is no time as constructed moments in succession.

Let me stress that, nowadays, we suppose that what happens in nature, like growth, the flow of rivers, and so forth, are seen as the process whereby objects and beings live their natural lives. Each and every one can be set to time---**but where is the time to set them by, since we do not accept that time is just there**?[183] They can be set to time only after an acceptable concept of time has been established with intelligence as applied to some repetitive cycles in conjunction with the sense of duration in the mind---in short, only after quantified time has been mechanised in a clock.

Thus one of the consequences of the space-time idea is that only mechanised time is true time for general application; all other semblances of time are just the chemical processes of things; but they are not useless in the scientific study of time because they can be set to time. The irony is that the attempt has been so successful that

---

[183] Even if it is just there we want to know what it is. But luckily, there are logical methods that give a clue how it came to be there. That is the purpose of this book.

sometimes we tend to believe that time is naturally in existence, and say, for instance, that the flow of a river is an example of time going; yes, but by whose time? Without time in the clock, the situation will be confusing, for nobody would know how much time is going. In a way the passage of existence and ageing is time going; but you have got to have the time already in a clock to know how it is going, and by how much.

The huge variety of objects in existence, each living its own life according to its chemical make-up, means that, although the growth of things may be seen as "time going", but that is not time for general use. I reserve the word 'time' for time in a clock, written in mathematics as ct. What that means is that it is time for an inertial frame, according to Einstein's theory of relativity. It means time requires points and mathematics for linking the internal sense of duration to external cycles that occur repetitively. Such a time is the creation of the human mind.

And here is the crucial question: following from the above, another puzzle arise, namely, can we apply our parochial time to events in the cosmos at large, say, regarding its past, present and future of bodies? How can we suppose that going round our tiny sun and calling it one year can be used as the yardstick to tell the age of the whole gigantic universe? Ten orbits is ten years but the period is so short because our sun is tiny.

And while we are discussing such matters, what about the definition of the time content of one year---or how long is one year, and how do we measure the length of one year in temporal terms? If we use distance it is the same as saying we can only know how time is passing and never what it is. For instance, as a matter of concern to all of us on this planet, when one year passes, you know you are aged one more year, that one year of your life is gone with it, but how much of your life span is gone---how do you measure that?

Thus, in my opinion, it is not very helpful asking how old is the universe, but 'How old is the universe by our time?' There is no other yardstick. On realising that the largest (or longest) unit of our time is the earth-year, which is just a measure of one orbit of the sun, estimating the age of the entire vast, gigantic and mysterious universe by this yardstick ceases to have any credible meaning.

Again, how at all does the cosmos figure in all this---that is, in the nature of our time? We believe we are here on our own as far as life is concern. The being of things offers no comradeship because everything is absolutely individual, except, perhaps, things like the branches of trees, where one thing depends on another. A universal time might give the impression that somebody is in charge and knows of us because he has given us part of general time, our version of it. But once our time is seen as uniquely our own (and completely secular) its range becomes doubtful when applied to the cosmos at large.

As is common knowledge, there are stars so immense that a million of our petty sun can find room in them. Given this fact, does anybody really believe that going round our tiny sun and calling it one year means we could determine the ages of events in the universe with the mere arithmetical accumulation of single units of our relatively short year? The abusive lunatics on the internet were wrong to attack me with personal insults; for I am only asking questions. There is no pretence that I can have answers to them. A thinker writes to invite discussion. Nobody who is no insane will write to dictate to the rest of mankind. Like the rest of us, I am also groping in the dark about everything. But we are now coming to the view that certain questions, since Einstein, undermine religion because the latter is not logically based. It is not the thinker's fault that it is not logically sound. However, it means we've been misled during many centuries of religious thought and obedience.

Even worse, apart from events, can we really seriously use these short years to determine the actual age of the universe itself? I have already said it is not correct to ask the question 'how old is the universe?' The proper question should be 'how old is the universe by our time?' Otherwise, by whose time, since age is related to time? The longest unit of our time is the earth-year, fifteen billions of which are mere fifteen billion orbits of the sun. We can think of acceleration that would give these 15 billion years in a short time! In any case what about the time before the sun came to be in existence? Furthermore, what of the time before the earth formed from interstellar debris into a planet and began to circle regularly round the sun, each of which is one year to us? Even this is not accurate enough. We should begin from when mankind acquired the facility to count the orbits of the sun as years, a most recent event by all accounts.

The religions have a lot to answer for. As infants or imbeciles, we are forced to worship anything in the name of God. It is coercion, and it is criminal. However, I am really surprised that the way and manner we get our years up to the centuries and even the millennia (merely by counting the passing years), has not undermined the so-called serious scientific theories of the age of the universe. These serious thinkers tell us that the age of the universe is between 13-15 billion years. Fifteen billion years for the age of a universe containing stars so immense that millions of our petty sun can find room in them; and especially when, in fact, one year is just the time for going round this tiny sun? I believe that this period is not even long enough for the formation of galaxies with their hundreds of billions of huge stars.

Frankly, the questions are endless. The age of the universe, in my opinion, is better left alone. It is not suitable for serious study, and it is not important anyway. The truth of the matter is that these studies cannot be justified on any grounds whatsoever. The objects are simply too far away to have any effects on our lives, the only sane reason for studying the cosmos in cosmology apart from vanity, is intellectual satisfaction. But astronomy is different; it is the subject we need to study very, very seriously.

We can legitimately study stellar events without worrying about the actual age of the universe; so, let us just say it is not amenable to human ageing concepts. I doubt that any inspiring theories will be missed by forgetting about the age of the universe. The universe is unimaginably vast; personally I shudder to think of its extent, nature, and how it came to be; these are human terms; they don't seem to apply to the cosmos. The mind boggles. And that is because we have forgotten the lesson from the nature of the quantum, which is that the terms we employ in our discussions of the universe and the sub-atomic world are human terms unknown in the universe outside a human mind or head. Some particles can appear in two places 'at the same time'. Yes, because they do not know what is 'the same time'. The universe does not respond to our time, concepts, categories or the things that pass through our minds.

Because philosophers are not properly appreciated by scientists, many momentous novelties introduced by Einstein (the reason they called him 'Philosopher/Scientist') are not given the necessary intellectual attention. Wittgenstein had much to do with that. At the time A.J. Ayer and Russell were stressing the importance of logic and philosophy, along came Wittgenstein to preach his logical mysticism to annoy scientists. Let me reveal a little of what has been missed by not paying attention to the rational thinkers whose only fault was failing to preach what people want to hear.[184]

Apart from our knowledge of bulky matter (or normal perception at the eye and sensory levels)[185], there is a hidden reality we cannot access at all, or can only partially reveal through mathematics and logic, even then in such theoretical formulas that very few of us are capable of following. After all the world has not been prepared to receive human beings in comfort; it would be different if it were so arranged. That hidden world is alien to us as we, also, with all our 'different-worldly-concepts' are unknown, even strange, there. These are different levels of existence. A simple illustration will make the idea clear: the golf and table tennis balls may look alike from a distance, especially if they have been deliberately equated in size. Only by examination can the differences between them be revealed. This trick of the eye is used a great deal in the Movie Industry, especially by Directors and Producers. The crooks also use it to deceive. But in theoretical physics mathematics can be used to sort things out. Yet the worst oddities still remain; or, to put it another way,

---

[184] Led by Bertrand Russell, they included Einstein, Professor A.N. Whitehead, A.J. Ayer, Professor Eddington and Karl Popper. Wittgenstein was first invited to join the group but was later booted out for his logical mysticism and the attempt to put an end to physics, as mentioned above.

[185] Bertrand Russell's book Our Knowledge of the External World launched the Philosophy of Science as a subject for academic study. Before that religion ruled the world in the academe and everywhere else; after that rationality should have taken over but, obviously, not everybody especially in the academe and religions respect Russell. The book is worth a serious study.

only a few of the oddities have so far been revealed. Much of nature is alien to us as we are to it. There are millions of things out there we don't know, some of which startle us from time to time. We first noticed this mystery during the development of the quantum theory and should have learnt the lesson well. There are things we cannot even learn until we are capable to learn them due to lack of cerebral competence or capacity, which may sound like a paradox, though it is truly the natural course of events in the cosmos, since we are always growing in brain capacity: an infant cannot learn to drive or fly an aircraft; the computer and internet were not discovered until recently, even though the facts, technical details and materials were there hidden in nature and waiting for discovery. It is not impossible that in future human beings can be compressed into a chip and transported to distant planets to be reassembled and live normally afterwards to colonise new worlds--- especially those whose surface gravels are all diamonds, if we're living under a plutocracy! It may be one way of moving to live on other planets, but at the moment we can't do so; and while the idea will be branded fantasy today, if and when it happens people will not even remember that it was once fantasy; they'll rather say it's the normal course of developments or progress.

We have to remember that a human being with all his massive brains, knowledge, and theories about the cosmos is physically less than a tiny drop of water in the Pacific Ocean by comparison to the universe, even to a galaxy of mere hundred billion stars. So far we have failed to make sense of it anyway; and it is about time to let it be. Astronomers can continue to skirt the fringes of some stars; yet again the nearest star is several light years away. I know of ingenious theories of rapid travel across space, but do we suppose that the human body can bear the strain of these velocities even if they were feasible---and for what purpose?

Humanity should always be understood as limited to the earth and its dwellers; as annoying as it may sometimes seem to be, your humanity and love is to another human being who can appreciate them, and there is nothing sweeter and morally inspiring than the appreciation of your fellow human beings. That is why many nations have 'Honour Systems'. It is about time we thought more about the world than the cosmos; there is nothing it can do for us whether we are rude or humble to it. It is a cruel world; no efforts should be spared in trying to minimise its harshness. That is the first lesson in humanity.

Time only became a major problem in science because of Einstein. As Professor Sir Arthur Eddington has remarked, it was not so before his researches about time. The scientific problem of time is different from that faced by philosophers. We have a situation where we are using time daily but cannot define what it is. Yet Einstein made time a separate co-ordinate in the determination of physical reality. That's what matters. The units of time remain the same; only the metaphysics of how we get them has changed.

The basic unit of time, the year, is virtually indefinable; while other units of time can be defined only in relation to the year as sub-units thereof. At the same time there are elementary conceptual difficulties in science about time: on the one hand, time is an artificial concept, called Space-Time, a 4-Dimensional metric of whose existence we have absolutely no evidence except that of imaginary 'thinkability' (as used by Einstein himself), or the Minkowski mathematics based on the square root of minus one ($\sqrt{-1}.ct...$) for time. This is ruled out as untenable in logic---if i is representing time as stated above by Einstein, no less, then what is the ct in the same equation representing? It certainly looks like an elementary logical error.

And yet, on the other hand, in all science we learn that time is naturally in existence, and always passing. But what is it that is always passing, if not the regular or repetitive cycles we use to reckon time? The cycles we use to reckon time of course go exactly with time, thus misleading us into thinking that the motions of the cycles are those of time itself.

We know they are not because time is discrete. It is based on the year and so it is bound to be discrete as the year is determinate, and passes by as the 'years' through replication not as a stream. Discrete time cannot run continuously like the cycles. So what is time? What does the actual physical nature of it look like? The natural aspect of time is the sense of duration in the mind. That is incontrovertible. Duration as time in the mind is always there, related to the sense of things enduring as part of the memory mechanism; for it takes time 'to endure'.

Some scientific thinkers have settled for a definition of time that equates time to being; that time is identical with 'Being' exactly, and that ageing is time going and taking life with it, not the other way round. The problem with that definition is the concept of points being required to apply to space to get the time---that is, create the time in the first place, once we agree with Russell that we construct our time under relativity. For how otherwise can we divide 'Being' to get time in units? We have to use repetitive cycles or motions. Also the idea of dividing time implies that it exists in some kind of a thread running through, yet the year upon which everything about time is based (related to or as fraction thereof), does not run through nature. It is one orbit---only one orbit of the sun and ends there. Our time is ended after one orbit of the sun. Another orbit is another year, not related to, or dependent in any way upon, the last year in the sense that it could not exist without the past year. Thus the idea of time running with 'all being' (as a universal time) has now been abandoned. That is another problem of time solved.[186] But it is interesting that all this goes back to the Russellian query after Einstein's demolition of absolute time: "If cosmic time is abandoned, what really is measured by the clock?"---in other words, how are we going to define time without the cosmic aspect of it? That is the question I have been trying to answer in this book. And I can reveal that it has engaged my attention for the past fifty years, enduring mockery, abuses and snobbery from the high and mighty in the academe and elsewhere! And now writers and thinkers have social media abuses too to contend with. If it gets worse we'll simply have to give up.

---

[186] But these problems are so many that no one writer can even list them all, let alone suggest solutions to more than a few.

CONCLUSION

As discussed above, there is no such thing as the physical passage of time.[187] As important as it looks, time is used merely to guide activities. For example, the idea of queuing up for anything is just a concept (or method) for getting things done among many people in an orderly fashion; it is not an independent entity to physically pass by on its own--- it is exactly the same with time as a concept or method for guiding activities among human beings. It does not exist outside the human mind. Everything that happens in the universe can be explained in terms of physical and chemical causes without time. All we need is careful analysis.

It is about time to liberate the human mind from the mystery of time. For time has influenced mankind so greatly that most of religion, all mysticism, magic and the black art, legends about the universe and even wizardry, together with all acts of unreason are related to time and the march of time as the story of history. In truth, history is the march of events, before and after, not time.[188] The dates of events are added to

---

[187] A second cannot pass without all of us living that second; whoever did not use that second to live did not exist. This is written even into the law. ('Where were you at ten O'clock?') Nobody can divest him or herself of time. The same principle applies to all other units of time. We live time. **In the absence of a universal time, we do not live by time, but live the time itself---we use it up as intervals between points; otherwise the time will not come to exist in sequences as Professor Whitehead has stated.** Yet, it must be stressed that life is not time (and time cannot be equated with life either), because time requires points. Time intervals are due to chemistry and other causes, making it quite impossible to equate with the life which supplies the points as time's basic ingredients. I should not be blamed for stressing this because it is the basis of our life---the most important idea in life overall. Solving the mystery of time is the closest we can get to the nature of life and intelligence.

them for purposes of **reference in the eyes of the law and cultural practice as regards evidence of existence and occurrences**. Otherwise, inheritance, succession, even the right to rule, etc. are all decided by preceding events, not by the number of times the earth has circled the sun, which merely means counting cycles and calling them 'years' or whatever. Again, everything can be explained in terms of chemistry or physics, the problem is that human beings dream, they also have imagination, and are fond of superstition and the secret desire to assume supernatural powers over their fellows; so they lie, cheat, and pretend to be what they are not. But thanks to Einstein, we can now demystify time, perhaps not completely, but enough to render it secular.

Again, time is unknown as a physical entity therefore it cannot pass by physically.[189] Otherwise what is the medium of passage?[190] The arrow of time notion is logically untenable without a medium, and physics has encountered no such thing in existence. Rather time consists of

---

[188] This is easy to prove since the events of yesterday, inevitably carried over, become today's events and the cause of what will happen in the future. This is how events cause history, not time. Moreover, while the future course of events can never be predicted accurately, the future of time can---the year is always the same. So if the time makes history we could predict it because we get all time as fractions of the year.

[189] It was never going to be easy to demystify time. It seems to cause everything. To show that it is not so was not going to be easy---some people do not even want to listen to the explanations. When they bring in consideration religion, life after death and time travel, then they (and some publishers) want to hang the theorist by the next lamp post! This makes Einstein the greatest thinker in human history for daring to question universal, absolute and general time.

[190] The general passage of existence notion is wrong because all existence is subject to time, yet not all are moving in the same direction: some are not moving at all, or moving and flying backwards, sideways, sinking and so forth. If time moves forward, how can it account for all that---millions upon millions of contradictory movements and activities?

mathematically created (culturally manageable) units of reality in human thoughts which replicate together with other factors in nature to pass by mentally[191]; the whole effects of these admixtures make it seem as if time itself is passing by, as the discrete units are expended one by one to create the illusion of 'an even flowing time through the cosmos'. All time happens as intervals between points: the shorter the intervals, the shorter the duration of time (in units)---seconds or minutes, even hours; and the longer the intervals the longer the duration of time, the years, for instance--- also in units.

As regards causality in the absence of a universal time, all we know is that chemistry, motion, force, etc are capable of giving man the sense of time 'as a period of waiting in the mind', duration, intervals between events, and points successively (something Leibniz pointed out hundreds of years ago); yet that is not time passing by physically. It is a mental note we make of the natural sequence of events when involved in any activity; so activity is the basis of time reckoning. What seems to be the physical passage of time are rather the cycles and motions we use to tract time (the years, months, weeks and days, all man-made), and which we count to estimate our own ages as well as the ages of other bodies in the universe. Needless to say, it is purely human, and, on the Platonic simile of caves idea, not the real thing---- only as we are conditioned to see it, especially since we cannot even tell the temporal length of one year! Otherwise, how long is one year without using any of the fractions of the year as a time unit to define it? But in case, just in case, we realise that we cannot even tell the temporal length of one year, then how accurate is using the year to

---

[191] This is the reason time is infinitely variable in the mind, in dreams, in the imagination or in art.

determine ages---especially the ages of the universe, of ourselves and of the planets? Concerning human beings, given the rate infants and children grow, at least we can guess what a year amounts to in biological growth---but what about the universe? Is there mathematics for this that I am not aware of? I doubt it. In any case, this is philosophy not mathematics.

Mathematics is not intellectually superior to philosophy, although Plato, Descartes, Russell, Whitehead, Eddington and others were right: philosophers need to know mathematics. But generally, philosophy is something like the sea of learning, while mathematics is the only safe means for sailing on this sea. However, if there's no sea there'd be no need to sail on it, or navigating it. Primitive people live without intricate, theoretical philosophy and the mathematics to go with it. But once there is a sea one has to know how to sail on it safely, professionally and competently---yet there has to be a sea, and we have many seas already. So the philosophers and mathematicians should work this out for themselves. I tend to think that philosophy versus mathematics is closely like the contrast between Plato and Aristotle: brilliant, mathematical, wise and a writer of genius; brilliant, sweeping, faulty, even silly in places, and yet a logician of genius. In our day, mathematics is vastly more important due to the rise of science; and philosophers unable to read much of science are not going to be able to contribute much of lasting value. Yet still, philosophy is the sea and without a sea there's no need for a boat. On the other hand, man cannot even think of constructed or structured time without mathematics; yet without that creating a civilisation is utterly impossible---how could we calculate the interests on loans from the loan-sharks, the IMF, Mortgage companies and the gnomes of Zurich?

Finally, what is 'the smooth flow of time' as against discrete time? Traditional time is supposed to be flowing smoothly through the universe or the world by way of the day and night system, growth, decay and any kind of decline and motion; it is always assumed to be there and does generally cover everything in the cosmos and the same everywhere---a second here being a second everywhere else. But once local time was discovered, thinkers realised that every locality can create its own 'local' time---but how? Thus the logicians set to work and discovered that all time is based on the yearly cycle but the year is determinate. It does not go on but ceases at a point and restart to orbit the sun again and again regularly. As such its fractions cannot be anything other than individual units in their own right---parts of the whole. Mathematicians then deduced that the second is the SI of time to confirm time as a discrete entity as required by cultural usage. Hence a second is a small bit of reality created with mathematics for cultural purposes; a minute is also a small bit of reality similarly created but a little longer than a second, and so forth, to all the other units of time. Each unit of time is part of reality created for cultural purposes out of the basic unit of time which is the regular yearly cycle. So giving any unit of your time to somebody is donating an irreplaceable part of your life away. This reality consisting of the world and all astronomy, plus the conditions they create for us to live in. So, for example, a very, very long, an infinitely long unit of time, would be getting nearer to the end of the life of the earth itself: thus a unit of time lasting four billion years would be a unit too far. This is important and must be emphasised: every second is a portion of reality in miniature, and all other units of time are either multiples or fractions of the second. As the SI of time, every unit of time consists of multiples of it, and it can also be pared down to the sub-seconds we know. And how did we deduce the SI of time? The answer is through the mathematical divisions of the orbit of the sun.

The conundrum is the connection between time and life, and I think life came first and that time is a method we have created to regulate our activities, especially as there can be no time in any part of the universe where there is no life. On this point I agree with Bertrand Russell that time is 'a construction' (see Mysticism & Logic, *ibid*), and consists of the intervals between points—as stated in The Analysis of Matter, *ibid*. He called it relation between points, but it is the same thing. That, of course, can be mechanised in the clock. All we need are the differences in lengths of intervals---seconds, minutes, hours and so forth---and time is completely demystified: a certain amount of space round the sun is called 'a second', a little longer and it is 'a minute', much more and it is 'ten minutes' and so forth---to all the units of time in use, since they have been created pragmatically. Thus we can answer Russell's question, 'if cosmic time is abandoned, what really is measured by the clock?' **It is existence converted to metric distances, and then subdivided from the year right down to seconds and again calculated in multiples and fractions of seconds as our manageable units of time. It may sound convoluted but that is what logical analysis of time reveals.**[192] **This is how our orbits of the sun is turned into units of time and continuous time: that one whole orbit is one year, subdivided down to**

---

[192] There is no doubt that this is the successful notion of time as a secular entity that changed all human concepts of existence. But one problem now is the habit of referring to video reversal of action as 'time going back' by people (including scientists) who are probably hoping that time travel will become 'a scientific possibility'. Properly all such videos should be called 'the review of recorded past events'. To regard video reversals as reality going back is so dumb that it does not require a response from serious thinkers. Why can't the whole world go backwards when we walk backwards? What is the difference?

seconds, sub-seconds and multiples of seconds---all units of time from the second are multiples of the second. Of course, the explanation of time was bound to be complex for time is very closely associated with life, the eternal mystery.

Otherwise the clock does not measure time. Time does not physically exist to be measured. The clock rather creates the time units out of the fabric of space, and **repetitively** because the whole contraption is based on the repetitive orbits of the earth round the sun. But since time is created from space it is right and proper to call it 'space-time', and wrong to try to equate it to space with mathematics based on imaginary time, as Minkowski did.[193] I may be wrong, but if the passage of time through nature, too, is its usage as discussed in this book, then I can think of no other mystery of it that logic cannot be used to solve.

Moreover, as if there're no philosophers in the world any more (or their thinking does not get through to people), presently what everybody generally refers to as 'time' in science, philosophy, religion and everything else, include mere gaps, pauses and intervals, delayed action, the duration of any activity, all kinds of motion, inactivity, gestures and even silence. Yet properly examined, these are all 'the natural ways' things happen due to individuation, multiplicity of actions and intermittence in all activities---human, animal and material.[194] Above all,

---

[193] Questions about time are so difficult that the Minkowski attempt was hailed as a work of genius; now we are told that Einstein didn't believe it or even bother to understand it.

[194] For an obvious example, if a person is playing football, his or her body is multi-tasked: running, shouting sometimes, breathing, kicking the ball, arms swinging in the air, and so forth. All of these can be set to time but they do not constitute time as we have in the clock. That is 'structured' or 'constructed' time, and Russell's query is 'where does the time in the clock comes from if not the

even scientists define time as 'just is', completely unable to tell what that is 'is'. The time we need to explain in science and philosophy is what Professor Whitehead defined as 'a sequence of non-interacting moments', or Bertrand Russell's 'constructed' time that covers everybody on the planet because, of course, it is based on (or produced by) the planet's own motions, and can be mechanised in the clock--- since the earth is always in motion--- to make it universally available, applicable and convenient, while showing how it passes by and seems continuous logically, even though, as fractions of the year, it consists of 'non-interacting units' which we know through science, philosophy and logic as 'seconds' or 'moments'

TWENTY-POINT SUMMARY OF CONTENTS

1. Following the discovery of local time, or t_1, Einstein declared that there are as many times in the universe as there are inertial frames, or planets. At once, logically time ceased to be general, divine or fixed. Instead it became 'a sequence of non-interacting moments' limited to planets--- according to Professor A.N. Whitehead. That is how time is conceived in this discourse. Those who are not aware of the new, secular definition of time are behind the times (It is regretted that some repetitions are unavoidable, given the mysterious and 'contorted' nature of time);

2. However, Bertrand Russell's response to Einstein was that there is no longer a universal time, which is treated above in two senses, (a) time is a human mental attribute and the universe

---

cosmos?'. That is called 'a theory of time' which the philosopher is expected to explain, because it is what we use to tell 'how long' any action takes or will take, etc., and is the most difficult question apart from the origin of life.

itself has no time or sense of time, being a bland, materialistic existence, where things happen through chemistry, random action or accidents, collisions or mergers; and (b) time is not general in the sense that a second here is a second everywhere else, for the universe is too vast and complex for one system of time to cover it all---there simply are far too many differing events;

3. After a century of discussing time along these lines, it is now generally agreed that (logically or religiously) we base all time reckoning on the yearly cycle, in so far as we use time in units. In other words, no matter what your beliefs are, once you use time in units, you're using it as a secular entity based on the yearly cycle.

4. This is because all units of time are fractions of the year.

5. So all time consists of units---being fractions of the determinate yearly cycle. The year ends and restart; it does not go on forever;

6. Mathematicians have worked out that, as a result, the second is our SI of time, making all other units of time, above or below the second, either fractions or multiples of the second. This is how we get the so-called 'most accurate clock' of the atomic time— i.e. based on the year and therefore also secular, though scientists make it sound as if it's the most miraculous thing in the cosmos. In fact, all time can be completely demystified. But the truth is that both scientists and religious believers seem to harbour the same supernatural ideas about time. Mankind creates most of the mysteries of time, otherwise it is logically not beyond the wits of man to explain---Einstein, Russell, Whitehead and Professor Eddington didn't think so.

7. Anyway (from 1-6 above) the logical trend makes time basically discrete and does not blanket the whole universe since discrete time---consisting of units---cannot flow or spread;

8. Also, discrete time can only be absorbed in units—in seconds or fractions and multiples of seconds known as units of time:

minutes, hours, days and so on, as we find culturally convenient. Actually, I think the interpretation of time should concentrate on the seconds, minutes and hours, or how we spend the day using time in units, as the weeks and months are not really relevant;

9. Discrete time, in units, would seem to pass by as the units are absorbed one by one;

10. This would make time in usage seem to pass by or disappear;

11. By this supposition there is no such thing as the physical passage of time; there is only the usage, the absorption of time unit by unit, as we do from the second to the hours;

12. Ergo, the usage of time is the passage of time, or the passage of time is the same as its usage. Time does not pass or go anywhere---it is used either for visible growth or the maintenance of existence, mostly the latter as people do not grow by the seconds and minutes. The passage of existence notion is wrong. Existence does not go anywhere: we live in the same country for all eternity. Growth and ageing in all nature are caused by chemistry and other factors, especially accidents and collisions. Time itself does not go anywhere or cause anything. It shows 'how long', and we also use it to live due to the passage of the years, which shows life moving on in its regular orbits of the sun and, rightly or wrongly, linked to ageing in the popular imagination.[195] (I am not at the moment considering the application of earth time to estimate the age of the cosmos. That's for the future.) But the repetitive cycles used to reckon time (the years pared down to seconds) are always moving, and they mislead us into thinking that time is in motion. Rather the moving cycles give us the time intervals to apply to reality so as to know 'how long' anything is or was there---how

---

[195] Every second is used to travel round the sun. It does not go anywhere, and certainly does not pass through. It's used by all of us on our journey and as we age incessantly both are assumed to be linked, probably wrongly. The same thing applies to every unit of time.

many cycles or years and so forth. Time has been completely misunderstood---see 17-20 below.

13. From the analysis above time and life appear to be inseparable, but the life came first because time requires points since it is obtained as intervals between points;

14. The Minkowski attempt to equate space to time was not successful---in any case, we now know that even Einstein did not accept it as true;

15. Since cosmic time is abolished due to the discovery of local time, time travel cannot be conceivable in any way. The clearest logic of all this is that there is no time at all. Human beings have, according to Russell, constructed time. Everything else that happens in the universe can be explained in terms of chemistry, etc.

16. Time can be obtained from any space, therefore it is correct to call it the product of space---as 'space-time', but not in the sense that space and time constitute one entity, which has become the source of scientific mysticism about time. Furthermore, the Minkowski ict equation upon which the theory is based ($\sqrt{-1}.ct...$) is flawed in a school-boyish fashion, namely, the square root of minus one is used to represent time and yet it is multiplied by ct which also means time. There are no two types of time in existence.

17. A time unit is reality reduced to culturally manageable units of duration with mathematics, making time a conceptual method to guide human activities, defined variously as a period of waiting, relation between points or intervals between points that we have cleverly found a way to mechanised in the clock, thanks to the mathematicians. The philosophical premise is that reality is spread out there forever and without limit; a minute of that is a clever contraption produced from space with points. Thus daylight is known to occur, but is divided into hours with mathematics---that is time in units, the twenty-four hours of the day by which we live. So the mathematicians who invented the

units of time are the creators of the world not God. All life depends on time sequences and they created them with mathematics. But philosophically there is no mystery because we count cycles for time and there are many things producing cycles in the world naturally. One orbit of the sun is one year, two are two years, and so forth. Then we merely subdivide the year into all the time units we have. We count cycles to generate the intervals we call time. The biggest cycle giving the greatest length of time (or interval) is the earth year, and that is what we rely on. There is no mystery only exceptional cleverness. As the process is continuous and metaphysically oppressive and unavoidable, the units of time (thus produced) became oppressively inescapable as if they're imposed from above by divine authority. That is how time came to acquire its counterfeit divine status to the glory of the churches. In reality, time is logically explicable in terms of mathematics and metaphysics, astronomy, chemistry and ordinary commonsense, roughly as I have tried to do here.

18. How was time originally invented, or to put it another way, how do we think the ancient thinkers and mathematicians conceived the notion that a time system is necessary for living in safety on this planet? My own (clearly) fallible supposition is that, the repeated astronomical and earthly feature became familiar as permanent features to everybody. The only thing missing was how long they endured, lasted or could be relied on to be there so that people could plan their lives by them; and it is supposed that after numerous attempts, the idea of using the Day&Night system together with the yearly cycle were regarded as the most reliable features to use for reckoning time and the units of time. After this it is assumed that, in the hands of mathematicians and astronomical speculators, the rest was straightforward---and, in fact, theories of time were still evolving as recently as during the Roman empire. So now sophisticated man knows that time is reality out there and each unit of time is a piece of that reality,

as each feature of that reality is known to last so many cycles that we have deliberately chosen for the purpose of reckoning time. We count the cycles as units of time. But in practice we use the long repetitive cycle of the year divided into fractions like days, nights, weeks, months, hours, minutes, seconds and the atomic units of time.[196]

19. In short, time is reality out there,[197] subdivided into hours, minutes and so on by means of cycles. The notable example (already mentioned) is how we've reduced the whole of the daylight period to twelve hours by using repetitive cycles. This applies to the whole of human existence: all reality, all units of time and all notions of time. Time is a contraption used to subdivide reality into culturally manageable units, as we cannot carry the whole of reality in our pockets---but the pocket watch can show bits of that reality regularly. Because this subdivision of reality into units of time is mathematical rather than physical, the passage of these units (as the passage of time) is also

---

[196] There is a tendency among scientists to regard the atomic time as mysterious, and also proving that time is universal because there are atoms everywhere and they too can produce time. This is wrong. Atomic time means sub-atomic pulses are used to measure the second more accurately, therefore it is part of the yearly cycle, or earth time of which the second is the SI. The last refuge of God is time, for almost everybody who can think, including scientists, believes that due to the mystery of time there is God. Unfortunately, we can use logic to give us a clear idea of time without God's involvement. For if there is God then he has made a colossal mistake about time because with the demise of the earth time will disappear---then what? We're often reminded that time operates even in the wild; but action in the wild are not caused by time intervals but by chemistry and other causes as mentioned above.

[197] Professor Whitehead's definition of time as 'contact' or 'instantaneous spread of the apparent world', was correct and very clever as it means 'perception' in the widest and deepest meaning of the word., even contacts in the womb are covered, yet at the time he wrote this the idea that contacts in the womb can be considered as perception had not even been discovered in psychology. He was far ahead.

mathematical rather than physical, proving that time is purely conceptual. The physical reality remains as it is only our concepts of it are altered in our heads or minds to guide our activities---e.g. it's twelve midnight and so one cannot go into the bush. Hence the two basic problems of time are solved, namely time is conceptual, and, being so, can only pass by conceptually; but time is closely allied to (and based on) reality to render it an accurate reflection of the outside world, so that roughly 31, 536,000 seconds will amount exactly to one year and we start another year. **Every time unit, even a second, is a piece of reality out there, since if the reality was not there the unit of time could not have been conceived at all.** This is how the world is organised successfully in purely secular terms---the creators of time invented civilisation and everything human, simply because everything human beings do depends on time. As I said before, it is the same story of Creation with the mathematician as God.

20. There are no physical gaps in the fabric of reality to accord with different units of time---e.g. the daylight is not divided into seconds, minutes, and hours. It is just there in its natural form. But we count cycles as units of time, and say the daylight has been there for two or three cycles, that is two or three hours, and so forth, to guide our activities within the daylight or by the daylight. Time is applied to reality to guide our activities. Time does not change 'the reality', rather our activities (and our time) are often altered to accord with the conditions of the outside world. But time, life and reality are so close together that we often confuse them, especially with our religious understanding of the nature of time. This aspect of time is important because that is what influences people's ideas about universal time, that it exists everywhere, even in the wild without anybody being present. Also people cannot understand how units of time are small pieces of reality created with mathematics, especially as reality out there is not physically divided. But people forget that

time is conceptual. There are twenty-four hours in a day, but there are no physical divisions of the day out there; it is conceptual. We carry our notion of time in the mind and apply to reality all over. According to the Einstein theory of frames, that is wrong. In the absence of a universal time, whatever one may use to reckon time is his or her own 'local time', not applicable elsewhere as Bertrand Russell as averred---earth time covers the world because we use the yearly orbit of the sun which covers the whole world's inhabitants.

LAST WORD

The mathematicians, who are almost always religious (because the subject is basically speculative, and speculative minds tend to wander), have been trying their best to make time the most mysterious thing in the universe, even more than the universe itself, with the concept of time dilation. But I suggest we ignore their intricate mathematics of time dilation and revert to how Lorentz defined it (and supported, in my experience, by the Mathematical Society of Japan.) Originally "Time Dilation" was called "the dilation of time as a measure of moving clocks". It meant that clocks moving fast seem to run slowly to people looking at them from outside the moving vehicle. Then, as Einstein lamented, the mathematicians took over and made it something 'we can no longer recognise'. All I can say is that thinking is difficult and easily led astray by people like tricksters, magicians and mathematical dreamers.

I believe time dilation helped Einstein to discover his theory of frames, because those moving with the clock never noticed any discrepancy except those looking at it from outside the moving vehicle----**meaning people in another frame. This is the interpretation we were given originally. Moreover clock dilation is not time dilation. Time can only dilate through changes in the motions of the earth. Clocks do not control all time; most of the time some clocks are not even working correctly! All those intricate mathematics are wasted because clock dilation cannot constitute time dilation.**

How did mankind jumped from this definition to the complicated mathematics of time dilation given by the mathematicians whereby time runs slowly with speed 'as predicted in Special Theory of Relativity, and since verified experimentally---though it only applies to very high speeds'? Yet at such speeds who can read any clock in the speeding vehicle? Meanwhile, everybody has forgotten what Einstein said about the Lorentz discovery. They like that; mathematicians, magicians and tricksters like to play down other people's tricks in order to promote their own, concentrate your attention on what they are doing. But this is what Einstein said: 'All that was needed was the insight that an auxiliary quantity introduced by H.A. Lorentz and denoted by him as 'local time' can be defined as "time", pure and simple'. It is this discovery of local time that led to secular time in the sense that everybody can create his or her own time---locally; therefore universal time does not exist. But because this is so crucially important in the fight against unreason, the mathematicians set to work to make it what it is not. All the difficulties I have experienced writing about secular time comes from this false interpretation of time dilation. And Einstein never predicted it; it was already known even before he started his work, and must have influenced him a great deal to discover his theory of frames.

This is poetic justice for me and all hardworking theorist bullied by mathematicians. For having rejected my technical papers a thousand times, I have finally caught the mathematicians with their pants down. The local time idea that came out of the theory of time dilation cannot have been predicted in Special Relativity because H.A. Lorentz said he did not discover Special Relativity because he failed to consider his own discovery of local time seriously---see Ch. 7 of Subtle is The Lord...by Abraham Pais, Oxford, 1982. For he had discovered it long before Einstein put pen to paper. Lorentz is rather said to have prepared the way for Einstein. Yet we are told that it was "...predicted by the special theory of relativity and since verified experimentally." These are all lies, and why should mathematicians peddle lies which they will use to reject other thinkers' submissions.[198] In fact, the local time idea was part of what inspired special relativity. Not only known but actually generated a huge debate. Einstein's subsequent researches about time was his greatest intellectual achievement. Russell was tentative, saying only that it was 'perhaps' one of his greatest. For me, there is no doubt because time is second only to life in all nature. Man often forgets that he has life and knowledge, and yet does not know where they came from and he will eventually die and they will disappear, where to, no one knows: from nothing to nothing but something in between. How do we assess that? We don't know. Metaphysically we are completely rootless. transient and lonely. I think those who preach love to one another are the wisest.

Meanwhile the legends about time are of course endless. What concerns me is that, in the era of secular time, thanks to Einstein's researches, as Professor Eddington has reminded us graciously, some thinkers have selected three bogeys to frighten us that time is still very mysterious, what for, no one can find out, but we should always remember that man is instinctively afraid of the dark---we just cannot know what may be lurking there. The fear of God due mainly to the fear of death as the total end of life is our basic torment---all of us. However here are the

---

[198] See the Penguin Dictionary of Mathematics, paperback, 2008, Ed.. David Nelson. Knowing what mathematicians can do to writers' careers and the advancement of knowledge, I think this matter should be investigated. Already we know there is widespread plagiarism, add deliberate falsifications to that and we're all doomed! I've always felt that crafty Einstein knew what he was doing when he warned mathematicians to leave him alone. After all, for what he did to our intellectual life, I believe he was the most extraordinary genius ever to come down in human flesh---if we get another he may come as a robot, having taken note of what happened to Einstein.

three latest bogeys about time: (1) Time dilation---answer, there is no time dilation because the earth's motions that we use to reckon time have not changed or **dilated; if true at all, it should be called 'clock dilation' not the dilation of the whole of time per se.** (2) Twins Paradox. This is an imaginary nonsense because the time dilation idea and the Minkowski formula upon which it is based are themselves fatally flawed. (3) Time travel. This is also based on the Minkowski formula that space and time constitute one entity and therefore as space curves (which is undoubted), it takes time with it, and this can be capable of enabling you to meet with your grand-parents before they're married. This is another piece of nonsense from lazy mathematicians because the formula is based on imaginary time coordinates which do not and cannot exist. With time alone there can be no imaginary time---how is it going to look like? In any case, as argued above, the Minkowski formula is logically flawed because he uses time twice in his ict equation, also coordinates cannot be transformed to infinity; imperceptible coordinates cannot be assumed to exist in support of any formula, particularly a serious one supporting life, like time.

There is a simple logical technique which shows that any clock's dilation for whatever reason is not the same thing as the dilation of all time universally.[199] It is this: in some tropical countries the sun goes

---

[199] Of course Bertrand Russell was right to deduce that there's no longer a universal time under relativity, in the sense that one system of time is covering the whole universe and the same everywhere---what was known as Newtonian absolute time before Einstein. Yet still there is a universal time in the sense that the time we have is based on the orbits of the sun and covers everybody on earth with only Zonal/Regional variations by the International Date Line. Basically time is the same as existence, or 'Being'; but existence is chemistry or caused by chemistry and the sense of time associated with it is personal---personal time. You cannot use that for work or anything else in society. It is your personal time

down after 6pm. If any clock has dilated anywhere (probably miles away), the regular sunset would not be affected in any way because the clock's dilation does not control time, or the regular time for the sun (normally) to go down. On the other hand, to argue that it is the effect of the clock's dilation on its holders that matters, will confirm the basic theory of local time, namely that time is your time (your local time) and not the time of everybody else---its dilation may matter to you but will not affect the regular time the sun must go down. And so in the time of 'sinners and unbelievers', only one clock has performed erratically (in any time dilation episode), not the whole of time per se. Therefore time dilation, even if true, is only the dilation of a single clock, and the mathematics which makes it universally applicable is based on a misunderstanding of time under relativity, namely, that it is

---

system. Therefore we have invented an all-embracing or constructed time for social interactions, work, science, philosophy and everything else involving other people. That all-embracing time is based on the repetitive motions of the earth. As personal time is limited to the self, the all-embracing time based on the earth's motions is also limited to the earth and its inhabitants, or dwellers. Primitive man confused the debate with the concept of absolute time, but gradually we're getting it right. But as everybody is aware, when the intricacies of time (of which there many) are further jumbled up with the amalgam of religion, fear of death, of god and the problems in life, the hereafter, mind games and even incidents from dreams thrown in, time can be made to look pretty weird even in scientific literature for all kinds of people to exploit to their own advantage. In mitigation He created Einstein, Russell, Eddington, Whitehead, Darwin and H.A Lorentz, whose combined suppositions make time secular, logically explicable and limited to this planet, with the proviso that other planets will have to invent their own systems called something else. In all probability, other ET civilisations, if there are any, can never have time systems similar to our own: Language and attitudes, bodily functions and the peculiar motions associated with them; they all help to condition our way of doing things and thereby affect our periodicities and mathematics upon which our time system is based.

no longer universal and the same everywhere. In particular it is this condition, that time is not the same everywhere, that mathematicians, being religiously inclined, seem incapable to understand. If time is the same anywhere then, of course, when a clock dilates here it will dilate every clock anywhere; but if any clock is only 'a local clock recording local time', then how could it dilate all time and all clocks? My conclusion is that time dilation is clock dilation. As such it cannot cause either the Twins Paradox or affect all time, since the all-embracing time under relativity, according to Russell, is a construction anyway.

Mathematics is not only important, it controls all science. But Bertrand Russell was right to advise philosophers to learn mathematics so as to avoid the mathematical dreamers and bullies, or at least stand up to them. For a start, every equation containing c^2 should always be viewed with suspicion in philosophy, because the speed of light is far beyond human capabilities, but to square it in any equation is approaching the realms of fantasy; on the other hand, s=ct implies that space has been equated to time, yet the Minkowski formula for that is fatally flawed---that is my last word.

However, I believe I have adduced sufficient logical arguments to establish that there is only one year, only one day, and only one second. The year is obviously one in nature; we start a new year at the end of the last one. The day, too, is only one, the sun never goes to sleep in darkness as we have to do, and the second is only one moment of consciousness (or of conscious existence), translated into a physical fraction of the duration needed to traverse the space round the sun by the earth.[200] So as our SI of time, the second is the means

---

[200] Everybody will agree it takes time to do so; well that is the total time allowed by the mechanics of the cosmos, and we take part of that. To mankind it appears

for transmitting the necessary periods of duration to all other units of time as multiples or fractions thereof---from sub-seconds to seconds, minutes, hours and so forth, as may be culturally necessary. Needless to say, all the units of our time can be changed, and tracing the origins of the sense of duration is utterly impossible, since it amounts to tracing the origins of the sun, to the galaxies and beyond. All we know is that the sun has given our earth a certain amount of duration to go round it, and our portion of that, as culturally feasible, begins with the second. But it means the right to live, to have a period of duration in which to flourish in this cosmos, cannot be investigated by human beings, since duration is caused by temporary events that may have since disappeared; but I believe it comes from random, non-sentient sources otherwise there would be sanctions attached to it with a purpose, and therefore it is either chemical or mechanical and purely material. What worries me is the origin of the brain and the nature of pain to our material bodies, especially since the brain itself never feels pain, does it regard pain as 'a malfunction', or 'a warning'?

---

as the sense of duration or being; so units of time contain specific amounts of duration, the gift of life, or being: one minute, ten, twenty, and so on. Sadly it is the essence of all our existence; we live by the time the earth is allowed for going round the sun. **This is the technical definition of conscious time, or how we metaphysically and mathematically convert space to units of time for cultural use---even for all eternity---and should be noted well because it is the ultimate logical definition of what time really is and whoever created it is the greatest philosopher of all time, not Plato. Anyway it was Einstein and Russell who set us thinking along these lines. It was bound to happen that man could eventually define conscious existence---this is it. Consciousness is the time (duration) and space allowed for breathing. Somebody will eventually link this to physiology.**

APPENDIX I

TIME AND QUANTIFIED TIME OR THE PASSAGE OF TIME

We are all fond of using the word 'time' loosely to refer to the passage of existence in any form whatsoever. That may be called 'the unscientific' notion of time. In logic, science and philosophy, however, time is what Professor Richard Feynman called 'how long we wait'. This translates into the concept of 'how much time', or quantified time, so as to be able to tell how long we wait in mathematical language for universal application.

In any serious discussion of time, it does not make sense to just mention time. It may take centuries to understand that 'Being' on its own is not time; it's true that you have to 'be' in existence to know how to apply points to create intervals of time between points, but being on its own is not time; being has got to do something to nature (that is know how to divide it into periodic intervals) to create time---that requires sentience. I insist that such a time system cannot move; what appears deceptively as the running of time consists of merely the motions of the cycles used for time.

As discussed above, motion is not time either.[201] It shows time going, but the time will have been created elsewhere beforehand. At best motion is silent time; but I think all that is chemistry, for chemical processing can impose (or require) a period of waiting at the visual level, which is the

---

[201] These are the two things (motion and 'Being') people normally assumed or imply to be time in ordinary conversations.

273

same thing as time. The truth is that all sorts of things may be called time but none can show how it began. We can only rely on how we experience it, and that is through the use of points as applied to space to create time intervals. Logically, this is the best we can do either in science or logic. Everything else is sheer religious humbug.

So motion is not and cannot create the time we can mechanise into the clock; the simple reason is that it is multitudinous. But we can count cyclical motions and call each cycle say, 'a year'. If this cycle is continuous we can have years and years in perpetuity; and a year of course is time; it allows us twelve months to do whatever we want to do; it is also the standard measure of age.

And that, precisely, is what we do to get the time to programme into the clock. However this can only show how much time is passing by means of our physical cycles and never the real thing. We count mere physical cycles and call them the rate of the passage of time---but what is the true nature of time? In my opinion time is the same thing we call life. The secret of life is the same as the secret of what we call time, with the proviso that all we can ever know of this time is how it is passing by; therefore the implication is that we can never discover the secrets of life either. Obviously we have to be in existence; and we are in existence (I think therefore I am!). So it means we are living (we're there or here) and choose to use a regular motion as the rate of the passage of time. The cycle does not call itself time. It's mankind that regards it as the rate of the passage of time, or the measure of duration to guide his actions. That's all there is of time. It is not as scary as was previously thought; it's merely a device to guide our actions. One can even tap the finger to the same effect. Let's say a thousand taps means an egg is cooked or done; that's not different from saying ten cycles (minutes) means an egg is cooked. The earth's orbit is so long that we've had to sub-divide it into smaller units of time. But every second is part of the yearly cycle, and therefore logically part of a cycle. A second to go is not yet a complete year.[202]

The nearest we can get to the definition of life is the logical definition of time, for the two are closely associated and don't seem to be separable. After many years of thinking it all came to me one day encapsulated into one word "When?" When is anything?[203] We can't have any existence without when (the time) it is or was in existence. Thus life is time and time is life, since we cannot define life without the "when?" it was or is there.

Ironically the logical definition of time reveals it as merely how it is passing by through the use of physical cycles. Thus a great conundrum, juggled round and round in the most exhaustive manner, becomes the greatest mystery. It is that time is life and life is time, simply because every second of existence is time; once you're alive you're expending time. It's not the same as saying it is the time allowed by God. It's slightly different; though I concede that the religions came close, very close. They've always had some back-room chaps bearing their intellectual burdens and some of them, like Johannes Kepler, were very

---

[202] Even nanoseconds are part of the earth-year. So are the atomic pulses used to mark time as they are based on the second---they merely lead to a more precise measurement of the second as a fraction of the earth-year.

[203] It's because existence implies time that the religions arrogated the right to decide what the nature of time was. Once time was liberated from religion, its logical definition was not difficult. The only problem is that existence still remains the attribute of time because we can only define existence in terms of the "when?" it's there.

good indeed. The difference between my theory and that of the religions is that I am asserting that time is life and life is time; on the other hand the religions claim that life spends on the time already allotted by God. Also, while they have no proof of their views, I have logic and Einstein on my side---with Bertrand Russell as a providential bonus! Time is life and life is time because you cannot define anybody's existence without when he or she was in existence; the quandary is that this time is one of our own creation, or construction. So, again, we have to theorise on the basis that man has a hand in the logical definition of existence. Thus the fields where man the observer cannot deal direct with reality are now three: Plato's simile-of-the-cave, the Einstein 3+1 formula, and our present definition of what we mean by existence---that nobody can exist without his when (time) of being there, but we construct this when ourselves out of the parameters we find in our environment. By the way, the Einstein 3+1 formula is included because of the time element---and we create the time---so it means we contribute to the nature of physical reality as perceived. That is to say, we determine physical reality from the three aspects of space and matter plus time, our time. This is the reason Minkowski incorporated the time element in his equation so that we can write S=CT to represent all physical reality and it is for the same reason of mathematical economy that mathematicians insist he is right. Unfortunately intention (or human desire) and physical reality are poles apart.

Again, my theory means your very existence is convertible to time the moment you are born not that the time is what is permitting your life to endure. The when of your existence comes in as soon as you're born and never stops---time is unavoidably continuous. Once a person is alive time takes over the control of his or her life in the following manner: you're alive. To continue to live on you've got to live strictly in accordance with the earth's motions and environmental conditions. These are what have been converted to time, based on the earth's motions. This time is unavoidably continuous because the earth never stands still. Thus from birth a person is controlled by the motions of the earth as we have converted them to time units---therefore time controls life. To be is to be part of the yearly cycle or be spending part of it---you cannot be without spending time. Nobody can exist without the 'when' or time of his or her existence. Life and time are inseparable. So to establish that time is secular is, for me, the greatest philosophical intuition or insight. I think it solves the last conundrum about time and life too.[204]

---

[204] The conditions conducive to life have been reduced to time units; these are based on the earth's motions and the earth never stands still, so every second is part of the earth's motions, and these seconds are oppressively continuous. So the when of a person's life is a time unit, without which he cannot live because it is tied up with the earth's habitable conditions, without which the earth would not be habitable. Thus if you do not live according to the seconds (or the motions of the earth) you'll perish. Life is therefore linked to time, or life is time and time is life. The whole theory is based on the fact that the earth's motions are continuous. It means you are free to live for one second because the earth's conditions have approved that it is save to live for just that second. With forward planning things are easier than that, but the principle is the same. You live because there is time for it allotted materially by the earth. If you have a second, a minute an hour or week to live (without medical conditions, it means if you are to live at all at any moment or time) it is because the earth has indicated that it is safe to do so, or that it'd last long enough for that; if it were not safe to do so the time won't be there, those time units won't occur. So we live by time allotted, but not in the religious sense. However, the religious chaps were very clever, as we arrive at the same conclusion by logical reasoning.

The Origin of Secular Time

Deciding 'how much time' by means of regular cycles is the main job of the interpreters of time. The context of any proposition (in science, mathematics and philosophy) must always show or imply the sense of 'how much time' in it, or expressly show the quantity of time proposed. Of course, time may pass when one is not conscious of it. But in all cases, when one wants to know how much time has passed, or will pass (as in futuristic propositions), mathematics must be used to quantify the time. And let me stress again that we quantify time by the use of external cycles in union with any sense of duration of anything whatsoever.

Quantified time is 'time in a clock', any clock at all. And the clock, any clock, can only show time as independent of space. Space-time is automatically quantified as it is derived from space with points, which is the only reason for calling it 'space-time'. Discrete time can only pass through the succession of the individual units. On this point, Leibniz was absolutely right when he said time is succession. What was lacking in his day was the concept of discrete time; with this new concept in our post-relativity world, we can now see clearly as to how time passes and seem continuous through the succession of its separate and individual units: second, second, second. Plus the hours, weeks and months all the way to the year, which also passes in the form of year after year after year.

It may seem surprising, the springs of a thousand legends, giving rise to supernatural speculations, that we have an extremely ingenuously smooth time system, so cleverly structured that it is there when we are born and there as we die, and always passing by. For this reason we know that "Time does not wait for anybody". Scrutinised under a logical gaze, however, time is not so rosy; it is only one moment, repeated to pass by and seem continuous so that arithmetic can be applied to its accumulations.[205] This, as we know well, happens when we reckon time for futuristic planning, and backwards as in historical narratives.

But for the union between the sense of duration and external cycles giving us units of time out of the moments of time, time for the clock would not exist at all. Presently philosophers see time as rather a straightforward pragmatic entity, albeit not as simple as it is normally supposed. It is partly a confidence trick that makes the clock work continuously, the trick of continuity is in the repetitions of the seconds, or of the units of time, all of which are to be understood as single moments---the realities---of quantified time. It is also partly physical (using physical cycles for the process of quantification); and partly philosophical, i.e. according to Einstein without time physical reality is indecipherable, or cannot be properly (accurately) determined, hence his equation for motion consists of the three spatial coordinates plus time in the 3+1 formula of physical reality. This, of course, is contrary to the Minkowski formula and I think this is much more scientific. It is true there is an element of subjectivity in it because the time is man-made, a human concept 'constructed' by man. But at least it is not as arbitrary as the Minkowski imaginary time coordinate.

To sum up, we have to recall that Einstein made man the observer part of the observed. Plato also made man part of the observed with his

---

[205] Let me explain that space-time is necessarily discrete. We have only recently come to understand space-time from Albert Einstein; yet time has always been discrete, consisting of only one unit (or moment) of time---of whatever length. For there is only one year, and all other units are obtained from the year in the form of separate units of time. To get more years we repeat the one year exactly. Thus we have second, second, second; or minute, minute and hours and so forth. Each is a moment (or a unit) of time in its own right.

simile-of-the-cave notion of perception, meaning we perceive the external world not as it really is but just how we are made to see it. When it comes to time (as the most important aspect of life) the situation is the same. We see the world and time not as they are but just how we are put together by the human architect to see it. Logic, of course, is our principal instrument of perception, theory and knowledge. Thus Bertrand Russell, as the great logician he was, summed the Einstein theory of time up and concluded that cosmic time should be abandoned since it cannot logically account for the nature of time as discovered in experiments. Professor Eddington also concluded that those expressing doubts about the Einstein theory of time were making meaningless noises. The founder of astrophysics was convinced by the secular theory of time. One reason, as I have pointed out, is that the religious notion of time and the nature of time discovered in logic are pretty similar: we don't know what it is but all of us accept and live by the yearly cycle as the passage of time, and have been doing so for centuries---centuries which are just the number of times the earth had circled the sun. Any good logician would sense that the true nature of time was not far away, especially after Russell asked the most important question about time--- if cosmic time is abandoned, then what is measured by the clock?

Unfortunately, researchers did not follow this logical trend to try and discover the true nature of time, but rather jumped on the Minkowski bandwagon to promote time travel. It's a sad reflection on the mentality of some writers that they should seek to twist the mind of mankind to concentrate on The Afterlife rather than the actual physical reality influencing human life.

I've always felt that if this had not happened (with numerous books about time travel selling millions while contrary suggestions are rejected), man could have done really good researches about the nature of time. Minkowski and Kurt Gödel bear the blame. But that is not all. Man is basically more interested in life after death than anything else. Well, if the Minkowski formula for equating space to time is not logically valid, it means it cannot happen, and if so then travelling by space-time is not feasible. We have to go back and research time as a secular entity that is separate from space exactly as Einstein made it in his special theory of relativity. At that stage the Russellian question comes up again---what is measured by the clock? Let us consider this question in the next chapter.

But I must stress that discussing the passage of time is necessary only to accord with popular ideas of time; otherwise I don't think time is ever in motion. To me the reality is that the cycles we use to reckon time make us think that time is running all through the cosmos. Yet time consists of separate moments, no matter how long each moments happens to be; so it can only advance through the replication of the units. Our time is based on the earth year which is so long that we've had to sub-divide it down to the seconds---but the main unit and its fractions advance by replication, not by running through the cosmos.

THE CLOCK MAKER'S PROBLEM

Pity the clock-maker. He thinks he is measuring time coming to him from the heavens; in fact, let me repeat what I said above that our time is created with our mathematics and is based on the repetitive orbits of the sun by the earth, and evidently the earth never stands still.[206] If ever it does stop going round the sun, our time system will be completely nullified; but, of course, life will go on. It is inconceivable that all life will be extinguished instantly the moment our time is (mathematically) nullified in that only quantified time would be lost. This is the best proof there is that life is not based on "time allowed", as the religions believe; rather time is a union between the sense of duration and external cycles---therefore man had something to do with the time we have in the clock, the only reliable time, as quantified time.

All the religions speak of "time allowed" for the duration of a man's life. They had to, because the nature of time is easier to explain as a providential bounty than anything else. To be honest, without a cosmic explanation for time, what is time; to put the question in another form, what is the origin and essential nature of time? Everybody believed that it's divine until Einstein and Lorentz found that it can begin from anywhere. Of course, it is assumed that the clock measures time. Even Bertrand Russell talked about measuring time, asking what is measured by the clock---but from where? And what is it that the clock measures?[207]

---

[206] The orbit of the sun is what we find most convenient to use as a measure of the passage of time. It's no accident. All the features of time are features of astronomy---events that happen to our planet and affect our lives. Hence the day and night system taught us how to keep time, the moon's phases and the seasons resulting from the earth's positions round the sun did the rest.

[207] Without the explanation that what the clock measures are cycles of duration, or duration reduced to cycles, metaphysically interpreted as a union between

The clock maker will say he invented the clock to reckon time in the sense that everybody knows---but what is that sense of time? The mechanics (or clock-makers) used the day and night system, the moon's phases and other astronomical features of the world as far as they're concerned, just to help us measure our version of a universal time. In other word, they're just as ignorant of the true nature of time as everybody else.

When it is postulated that general time permeating the whole cosmos (and therefore the same everywhere) does not exist, the first implication is that every 'body' (or planet) has to have its own time; it is not coming from the cosmos therefore it must have originated on this planet. So let's find out how it all began. That is the first implication. The second is that, as a result, cosmic time is abolished---although it sounds tautological, it still has to be emphasised, as well, and most clearly because the 'cosmic time instinct' is permanently ingrained in the human mind. One reason is that time cannot be suspended; but the more cogent reason is sheer intellectual incompetence plus fear of the unknown. We are always using it, and so it does not make sense to just say that it is not there. But if it is there, and did not come from the cosmos, how did it begin? And the obvious fact is that it is always there. Even before we are born, and also as we die to leave it behind. Yet it

---

duration and its conversion to external cycles, time can never be logically accounted for. We will just go on using it---but in what form? In the form of units (year after year after year, and all the seconds and so forth derived from the year); yet that means the same thing, namely, a union between duration and its conversion to external cycles. For the year is only a physical orbit of the sun. It is not time. It is the practice of humankind to call it 'a year'. We use it as our basic unit of time, as a matter of convenience. Otherwise in nature it is not time. As a matter of fact, we can use something else---we can tap the finger, for instance.

The Origin of Secular Time

cannot be supposed that each body's time is a version of something 'naturally existing', whether it permeates the whole cosmos or not, with the necessary but illogical (little 'academic') proviso that it may not be the same everywhere but varies with individual bodies in accordance with unknown natural laws.

It is plainly evident that this erroneous sense of time dominates scientific thought. Hence time is not defined in physics; and as a result, the Minkowski fiction makes sense to some scientists. They just say "as time goes by". Only Professor Arthur Eddington has redeemed physics by warning that it must never be forgotten that the 4-D geometry formula is "fictitious and arbitrary"---but they have chosen to ignore him, partly because Eddington was afraid to mention Minkowski by name, or maybe he's just cunning. The era was incredibly sensitive: There was Einstein, Planck, Russell, Whitehead, and the mass of aggressive no-nonsense mathematicians who regarded Minkowski as the genius who made relativity accessible to scientists. Thus Eddington had good reason to be cautious---nevertheless, since then everybody refers to the concept of space-time as 'artificial'. Let me explain another small point about original ideas. Even the originators do not stick their heads on them, because they're never absolutely certain that contrary ideas would not emerge; and we all know that most of the time they've emerged to shame cocky theorists. So even Eddington might have been a wee bit afraid of the pure mathematicians---even Newton was, and if David Hilbert is to be believed, then Einstein too was!

Thus Russell's query is important, namely, "If cosmic time is abandoned, what is really measured by a clock...?"[208] My answer, of course, is that

---

[208] ABC of Relativity, Ch. 4.

outside the union between the sense of duration and its conversion to external cycles, time does not exist to be measured.[209] The very act of 'measuring' is the time in essence---like moving from point to another point, time is going, so that time becomes 'relation between points', or intervals between points. The cycles give us the sense of time, of the sense of waiting or time units (the years, for instance), and the time units constitute the time: a year is only a physical cycle, yet it is also our time, the basic unit out of which all other units are derived. How the mathematicians did this I do not know---try and error, I suppose!

However, the cycles are the deliberate creation of man for the sole purpose of converting the sense of duration (of anything or any event, like the period it will take to reach the village from the farm before nightfall to avoid predators), to his time units to guide his activities. So the clock does not measure time; it rather reproduces units of time programmed into it by the clock-makers. It should be remembered that the seconds are put there by the clockmaker; but where do they come from? The answer is that they come from the subdivisions of the year. Otherwise the time does not exist anywhere to be measured---the units constitute the time. Without the year there will be no seconds, and the like, all of which are derived as subdivisions of the year. As hinted above, you can even dispense with the year and its subdivisions and tap your finger, if you will not get tired. A million taps means it is time to go to bed, and so forth; outside the units of time, time does not exist to be measured; but the units are the creations of man as quantified time to

---

[209] I believe the origin of the sense of duration is in the brain; that it arose from how the brain was put together---generating the sense of time-lapse---and grew with it. For the brain was not formed at once; there must have been lapses of time and that created an instinct buried deep in the brain and it affects us. This is only speculation but gives us something to think about.

record the passage of existence in manageable units for cultural purposes.

I conclude that what we call time is the mere physical manifestations of it that we use (as periods of waiting) to organise our lives. These manifestations (or physical cycles) that we call 'time' are in motion of course: we count the earth's orbits---caused by motion---as 'years'. That process has been taken as the march of time. But we don't know what time is to tell whether it is marching or not. My feeling is that events do march and they have time associated with them thus misleading us into thinking that it is the march of time. In fact we can never know what it is, if it exists at all. The causes of the cycles we call time units may be chemical, inertia, dark matter, momentum, motion, kinetics, delayed-reactions and so forth. They cause what we experience as time. If real time does exist we cannot know it because it is shielded by the parameters we use for time. This echoes the Platonic Simile-of-the-cave again---it seems man just cannot perceive real reality. To me that's not so strange, for we are so insignificant anyway. The real surprise is the incisive power of our brains.

APPENDIX II

THE PRINCIPLE OF MATHEMATICAL EQUIVALENCE

In nature there is reality and our perception of it. I subscribe to the Platonic simile-of-the-cave theory of perception. In the word 'perception' everything man does in life is implied, including mathematics, since we can only act by perceiving the true nature of the physical world; I am using the word in a sense akin to 'experience'. The problem is pure mathematicians normally are permitted to imagine things to satisfy their nostrums, so that they do not rely on their percepts alone. However outrageous, they can defy reality, even gravity, logic and common sense, and leave it to the applied mathematicians, to find out whether what they have assumed is really there in nature, so that their theories based on them can be seen as true or not. In no other profession is this sort of thing allowed. Even one of the greatest mathematicians Britain has ever produced, Professor Sir Arthur Eddington, criticised that common mathematical tendency in his book, The Mathematical Theory of Relativity. I have quoted him above in the text, but it will do no harm to repeat it as it is vitally relevant here. He said: "The pure mathematician deals with ideal quantities defined as having the properties which he deliberately assigns to them. But in an experimental science we have to discover properties not to assign them…" The principle of mathematical equivalence should make them think of the practical consequences of their imaginary properties, although I doubt it, but that is another matter. The rule is that mathematicians should not seek to make the basic features of nature what they are not quantitatively, or cannot be physically; any such

287

propositions are bound to falter. Note that we are talking only of basic phenomena. By the very nature of man, it seems everybody can make qualitative/physical changes in peripheral nature not quantitative changes in the fundamental aspects of nature, and time is the second most fundamental feature of both nature and life.

The principle means that, in effect, one cannot use mathematics to state, say, that there are ten trees in a field, and propound theories about them if, in actual fact, there are only two. This is slightly different from assigning imaginary properties to nature. It is different because it relates to 'quantities'. Six into four won't go, or something like that. The principle of mathematical equivalence rules that, to accord with physical reality, one can only talk about two trees, or as things are not as the mathematicians want them to be. Nature is not there for the convenience of mathematicians; it is neutral. That was the advantage we gained when the ancient teleological interpretations of phenomena was discredited. Therefore this rule is not to be scoffed at. I regard it as one of the strictest doctrines in logic, metaphysics and science. Science means logical thought in physical applications; metaphysics is logical thought in abstraction and mathematics is logical thought by means of symbols rather than language to facilitate the handling of size, weight, distance, volumes and complexities.

It is not often realised how progressive is the study of philosophy. Quietly but surely, many entrenched myths from our primitive past are being discredited one by one by philosophers. One of them is teleological argument. With that and many other ludicrous intellectual fashions out of the way, it is unacceptable to regard any concept as 'compounded for the convenience of the mathematician', as Russell defined the Minkowski theory of space-time. Someday, we may get scholars writing about the many myths philosophers have discredited through their quiet researches to foster science and progress generally. So I regard this principle of mathematical equivalence as a strict and necessary doctrine to prevent mathematicians arrogating the power and right to alter nature quantitatively in the fundamentals of physical reality. We shall, and should, continue to alter nature qualitatively to our benefit---gardens, buildings, roads, cities, waterways, canals, railways, bridges, tunnels, all science (bar destructive devices), and all art, sports and so forth. They do not change nature but beautify it; but quantitatively, never. We cannot make one object two, or two objects one, physically. It is not possible realistically. Not in reality only in the imagination; to jump from the imagination to live conditions can be dangerous.

The origin of the rule will help the reader to understand it well when spelt out: it occurred to me when I was pondering Hermann Minkowski's claim to have made time and space into one entity as from the moment he outlined his theory, as previously quoted, in the following outrageous (even cheeky) statement: "The views of space and time which I wish to lay before you have sprung from the soil of experimental physics, and therein lies their strength. They are radical. Henceforth [that is, from the moment of his lecture] space by itself, and time by itself, are doomed to fade away into mere shadows, and only a kind of union of the two will preserve an independent reality". This is to combine two things in nature into one with mathematics ('a kind of union of the two...') It means he knew they were two independent aspects of nature. How could he have made them one from the very moment of his lecture? (Yet mathematicians continue to accept his formula as true.)

The Origin of Secular Time

He spoke of experimental physics. In fact, the only experimental evidence pointed to time being 'local' in nature; and Einstein adopted it in his special relativity; it's the Lorentz t^1. There was no suggestion that time had been found to be inextricably intertwined with space---rather the suggestion was that time could not be had without space; and that once you have space, you can create your own local time. What Einstein did was to interpret local time to mean "The only Time" we can have.

The actual physical reality known to be in existence was precisely as Minkowski himself stated it---namely, that time and space were two separate things. But it is interesting that he sought refuge in experimental physics. In that sense he did not breach the principle of mathematical equivalence. It shows that he was really a very good thinker; he had to be that good to convince Einstein to adopt his formula for general relativity, which came ten years later. The unfortunate thing for Minkowski and his followers is that the evidence he cited was really irrelevant to the claim he was making. He needed physical support that time and space are inextricably intertwined and therefore constitute one entity. The evidence that had been discovered by Lorentz and Einstein was that time was essentially local in nature, leading to the supposition that 'there are as many times as there are bodies', and that, additionally, time is different in different places, and also under different conditions. The principle of mathematical equivalence can be used to refute Minkowski's claim to have made them into one entity as from the moment of his lecture.

The rule stipulates that he could only have spoken about time and space as they actually were in physical reality, which, he admitted, were two separate entities. The reality before Minkowski was that there was space, and there was time. Even the great Einstein himself made them independent in his special theory of relativity. So it did not surprised me that Professor Sir Arthur Eddington and Bertrand Russell described the Minkowski proposal as arbitrary and fictitious. However, it did surprise me that mathematicians ignored this strong condemnation to claim that they could not understand Einstein's ideas without the Minkowski fiction.

That made me sit up and think, think of a principle to require mathematicians to relate their suppositions to exactly the nature of physical reality laid out before them, not as they would wish it to be to accord with their nostrums. I came to the conclusion that mathematics can only mirror reality, not to alter it with mathematics alone. So the principle of mathematical equivalence is this: Mathematical statements (or equations) must strictly accord with physical reality. That is the true meaning of the term "equation". It means no mathematical quantity can exceed or reduce what the actual physical quantity is. No mathematics can make one thing two, or two things one, without physical divisions and unions. Minkowski failed because, as Professor A.N. Whitehead has pointed out, time and space still pass through nature as two entities, not one. Of course, Professor Whitehead did not know what we know now, namely, that time does not actually pass by physically, but only by means of the procession of its units as created by man with his mathematics out of the orbits of the sun. Let me remind the reader once again that the theory is that the sun gives us our time units physically (otherwise we couldn't create the SI of time with units that multiply to coincide exactly with a full orbit of the sun). As these units of time proceed successively, the time is passing by. Since time's intrinsic nature remains unknown, the passage of time cannot be explained in any other manner in logic or scientific thought; but the SI works so it must have something to do with the true nature of time, surely? This is not fantasy.

The Origin of Secular Time

The real irony of the situation (which makes me smile to myself happily in self-satisfaction), is that the mathematicians tend to be religious; yet the creators of the SI of time helped to accelerate the demise of religious ideas about all time, the end of time, end of the world, and even of the universe, because by the mathematical SI of time, time became physically based, secular and traceable to scientific thought.

Yet, despite all that, the serious matter is that Minkowski rules the world from his grave, and I think that is distorting theoretical physics, even though mathematicians take mischievous pride in the situation out of spite for philosophers and logicians. The term 'space-time' is everywhere taken to mean space has been equated to time or vice versa, generally meaning space and time constitute one entity as Minkowski proposed after relativity and that the whole momentous 'creation' was achieved with mathematics---which would at once breach the Principle of Mathematical Equivalence. In other words, mere mathematical symbols ("S=CT...") are said to have reconstituted the whole of physical reality. The reader must agree that if this is not logically accurate, then theoretical physics is being adversely affected. So I must be forgiven for repeating the whole debate again. After he was persuaded or coerced to adopt the Minkowski proposals, Einstein wrote in his seminal book RELATIVITY (Routledge edition, 2001, Part One, Sect. 17, pp 56-58): "We must replace the usual time coordinate by an imaginary magnitude $\sqrt{-1}.ct$ proportional to it..." To justify this in metaphysics or epistemology, Einstein wrote (pp56-57 of the same book), since, as I have stressed, mathematics cannot alter reality but only reflect it, even the great Einstein wrote, "...for in every event there are as many 'neighbouring' events (realised or at least thinkable) as we care to choose...", **and that, for all I know, is the crucial point**. It amounts to stretching the transformation of Coordinates to infinity.

Is that acceptable for the determination of the nature of physical reality? Can we rely on human 'thinkability' to tell us the true nature of the world? To me the obvious answer is no. Therefore space cannot be equated to time or time to space, so the equation "S=ct..." is flawed not only in logic but also in the nature of physical reality it portrays, since time is conceptual and not physical or part of physical reality; that is why we can't find how it physically passes by. Yet it is true that space and time are so closely bound up together that we have no means of separating them beyond what I have been trying to sketch in this book. Thus to call them 'space-time' is correct in the secondary sense that they are virtually inseparable.

It means space and time, though independent, cannot be experienced separately for whatever metaphysical reasons, and so they come together in the perspective of man (the observer); and that man affects how human beings perceive reality, as Plato suggested. This is the ultimate of metaphysical reality and there is nothing we can do about it. Certain things in the world can never be explained. All we know is that we are alive because of it, and thank God for that. Of course, probably God had nothing to do with it at all. It may very well be the case that because of our convoluted way of getting our time in units from the determinate yearly cycle, it is man who had need of space in getting his time: we cannot get time in units without using points, and using points enforce the need for space. Without the yearly cycle man could never have invented a time system, because time is strictly linked to the environmental conditions created by the yearly orbits of the sun by the earth together with its rotations. That is why twelve midnight is not the time to go to the bush.

The Origin of Secular Time

However to merge them in certain mathematical equations, where necessary, to be logically accurate, we must revert to the equation 'S+ct...' otherwise known as the 3+1 formula. But of course mathematicians will never do that just to please philosophers, and yet nothing will go wrong because time is always the same, what is sacrificed is the true nature of reality. There is no doubt that man affects how we see reality, because time by which we do and know everything is human in origin. To invent clever mathematics to try to demonstrate that it is not so is to distort reality.

APPENDIX III

WHY SPACE ON ITS OWN IS NOT "SPACE-TIME"

In Einstein's special theory of relativity, we learn that, "In the absence of gravity, space and time are distinct entities. In the metric of special relativity they play distinctive roles."[210] Nothing in special relativity has changed since then to make all space "space-time". Yet in all their suppositions cosmologists and astronomers always refer to space as space-time.

Let me set out the facts as they are at present, as argued all through this book, and hope they will see the light. To begin from the very beginning, the whole (contentious) debate about space and time began with the work of H.A. Lorentz; until then space was space and time was time. It is true that in special relativity Einstein made space and time dynamic rather than the Newtonian absolute; but being dynamic merely means they are changeable under different conditions. But about time alone Einstein avers that he was able to complete special theory of relativity five weeks after he gained the insight that the Lorentz idea of 'local time' can be defined as 'time, pure and simple'. So let us examine the Lorentz notion of local time.

H.A. Lorentz found that time runs slower when in motion, known as "the dilation of time as a measure of moving clocks". He could not understand why and literally put it aside. He called it 'local time' or t^1. To

---

[210] Professor Jeremy Bernstein, in *ALBERT EINSTEIN: and The Frontiers of Physics*, Oxford, 1996, p110.

him it was not 'the true time' but a mathematical auxiliary or curiosity---not very important. Time, he said, was time, denoted with t, and t^1 was something you get as your local time, but certainly not applicable in the outside world as time, because it was a mere mathematical curiosity. May I remind the reader that all this has been given in detail in the text above? I have even mentioned Lorentz's own statement that he thought he failed to discover special relativity because he did not regard time dilation as important.

Strangely, however, as one of his brain waves, Einstein worked this into his theory of frames. The dilated time was 'local time'---the time of your locality. Now, if the universe was fragmented, then local time would be somebody's time, which to him would be running normally like any other time, but to outsiders would be running erratically (or slowly, in this case.)

In actual fact, that was the case with the Lorentz discovery. People outside the moving clock saw it as running slowly; but those carrying it in the moving vehicle noticed no difference in its performance. That is the genesis of the Einstein theory of frames. Otherwise time was separate from space. What you will find is that it varies under different conditions, simply because everybody has to have his own 'local time' in his locality or inertial frame. But since time is continuous, and having made it a separate co-ordinate in the study of phenomena, dynamic space would have different time co-ordinates at every turn. We recall that Bertrand Russell has stated that from the sun's point of view the tram never repeats a former journey---because the time co-ordinates would be different. Since time is a separate co-ordinate in the determination of physical reality, different time co-ordinate implies a different situation, different physical reality.

This was the state of affairs when Hermann Minkowski came in with his theory of 4-D geometry making time part and parcel of space---all space. So that cosmologists and astronomers call his theory "The Minkowski Universe", meaning that all nature is subject to the 4-D geometry, where time and space constitute one entity. But let us swiftly add that the foremost mathematical interpreter of relativity was our own Professor Sir Arthur Eddington, the man who confirmed the general theory of relativity. He wrote the definitive book on relativity, called The Mathematical Theory of Relativity. About the Minkowski 4-D Geometry, he stated clearly on Page 9 (Ch. 1.1.), as already quoted, "Such a mesh-system is of great utility and convenience in describing phenomena, and we shall continue to employ it; but we must endeavour not to lose sight of its fictitious and arbitrary nature."[211] He was not the only great mathematician who described the Minkowski formula as arbitrary. Bertrand Russell also said it was based on an arbitrary assumption. He made it plain that because of that the derivation of the Minkowski 'interval' as time from space was not valid.

Let me try and explain again the reason mathematicians still adore the Minkowski theory---even though they know that it is fictitious. It makes things easy for them. Yet it is not true. They accept the novel Einstein notion that time must be made a distinct co-ordinate in the description of phenomena. The problem is that at the same time Einstein made all time (any sort of time) 'local time'---the time you create for your own local purposes, as Lorentz had discovered. Einstein extended the Lorentz idea to all nature. With the universe being fragmented, it was impossible

---

[211] The emphasis is mine. I have had to mention this several times, because, quite honestly, I am outraged by the mathematicians' desire to perpetuate the Minkowski formula as if it is really true of physical reality---yet it is not, and they know it. At least one of their own numbers told them so.

that one system of 'dynamic time' (as opposed to 'absolute time'), could apply with equal validity to all fragments of the universe. As a result he said there are as many times as there are bodies in the universe. Nobody can contradict Einstein on this matter. But mathematicians found that creating your own time to add to phenomena to acquire concepts of physical reality puts too much power in the hands of mankind. (I suspect there are religious sentiments in this.)[212] Besides, it was complicated. The Minkowski system was easier;[213] you just have to mention the Minkowski space or ds² and move on. It comes with time already embedded in space as part of it---so the whole of space is 'space-time' and every time is also 'space-time'. The caveat of Professor Eddington was quietly ignored. Soon everybody forgot about this; Eddington and Bertrand Russell were dead; and there was nobody clever enough to notice the discrepancy and question them about it. Of course, that leads to a distortion of relativity, but mathematicians are the arbiters of truth in mathematical physics and they were the ones benefiting from the Minkowski theory, and therefore preserved it. Otherwise it is not true that all space is 'space-time', while all time is also 'space-time'.

---

[212] The Minkowski formula makes time universal again after Einstein namely, as something in general existence mysteriously (harking back to Pythagorean mysticism in mathematics), which can be invoked with the appropriate mathematical symbols; not as something you create in your own local space with the application of points to space, which makes time completely secular. It seems to me that humankind is not ready to accept time as purely secular. Those of us who have already made the necessary psychological adjustments for accepting time as plainly secular are not regarded as normal.

[213] It was difficult in mathematics but easy in logic and philosophy; and let me hurry to add that, because of the involvement of time, the whole notion of local time or space-time has philosophical implications, since time is the second most important thing in the world, second only to life itself.

Yet it is true that time is always space time. You cannot have time without space; not because the space comes with time inside already, but because all time is known and used in units and units only, which can only be had by the application of points to space to create the time intervals as "relation between points". There are elements of time in the mind as the internal sense of time, known as the sense of duration. But we have got to link duration to external cycles to give us usable time in units, as I have explained above. For example, without space we cannot have the year; yet the year is our basic unit of time out of which all other units are derived. This brings a little complication but nothing serious. The reason is that you can only create time, as 'intervals', or as 'time units', as I suppose (because the year is only one unit of time and we derive all other units from the sub-divisions of the year with points or mathematics), with the application of points to space, thus making time a product of space, and therefore 'space-time'. The truth of the matter is that you cannot have time without using points to divide space; it makes time necessarily discrete, being the product of points. Therefore time is always 'space-time, or properly 'space-timed'. But that is all the connection between space and time, except that space is required, again, for displaying time in units as we have in the clock.[214] The clock,

---

[214] The poignant question posed by Bertrand Russell comes up again, namely, in the absence of universal time, what really is measured by the clock? (ABC of Relativity, Ch.4.) This is a very serious matter, because if cosmic time is abandoned, there is no time, or any logical explanation for the time we have. The answer, of course, is that the clock does not measure time. It is deliberately programmed to *reproduce* specific units of time: second, second, second, leading to minutes and so forth, to accord with the cycles of the earth, so that about 31,536,000 (or so many) seconds will coincide exactly with the earth's orbit of the sun, called 'one year'. To have more years, we go round the sun again and again and again---hence perpetual time. Units of time in procession give us

any clock, does not give 'flowing time'. It merely reproduces units of time programmed into it. The old mechanical clock based on coiled springs gave the best illustration. The springs are manufactured to release units of time: second, second, second. If one failed to rewind the springs, the clock stopped ticking. The springs provided the clock's energy, but were strictly programmed to reproduce time in specific units only.

After the time is derived in this way, it becomes separate from both the space and the points used in creating it. That is why Einstein made them separate entities in special relativity. For, apart from the condemnation of the Minkowski 4-D geometry which assumes that time and space constitute one entity by Russell and Eddington, Professor A. N. Whitehead has also pointed out that time and space still pass through nature separately---not as one entity. To add to these, I have humbly suggested the Principle of Mathematical Equivalence above, which can also be used to denounce the Minkowski arbitrary and fictitious formula.

---

continuous time. From the Einstein concept of space-time we know that time, since it is produced with points, has got to be wholly discrete.

APPENDIX IV

THE MISCONCEPTIONS OF TIME IN RELATIVITY

It must not be supposed that the problem of time in relativity has been conclusively settled. Relativity is physics. When a problem is solved in physics the solution is always clear, precise in mathematics, and universally applicable; but time in relativity at present is very vague, neither definite nor precise, not least because consideration of time is a philosophical inquiry, and a very serious one too.

The arguments here are that the original Einstein theory of time can be used to solve the passage and continuity of time. Unfortunately, Herman Minkowski made the question of time in relativity immensely complex and vague, not at all like the original notion proposed by Einstein. Indeed, as a result, the question of time on the whole is destine to keep the philosophers busy for several centuries as their nostrums become footnotes to Einstein instead of Plato. As regards the physicists and cosmologists, as opposed to the philosophers, they believe that the Minkowski theory makes things easy for them; the problem is that it is just not true of the physical world.

Bertrand Russell has said the concept of space-time is perhaps the most important theory Einstein introduced. To me, there is no doubt (no 'perhaps') about it. It is the most revolutionary theory in human history simply because time is second in importance only to life itself---and yet that life cannot even be lived as a well-organised existence without time. That is how momentous time is in human affairs; and Einstein has shown that it is very different from what it has been traditionally assumed to be. Secondly, he insisted that it should be taken as a separate coordinate in the study of phenomena. In the determination of physical reality, because of Einstein time is a co-ordinate in its own right just like the height, length and width of space are, thus making Man, the observer, part of the observed, since he has to add the time in the 3+1 formula. Those mathematicians who assume, on the Minkowski theory, that time can be incorporated into space with mere mathematics so that we can dispense with the 3+1 formula and the metaphysical role of man in the determination of physical reality, are contradicting Einstein, which is something approaching a hanging offence in science. On the contrary, it is possible that the passage and continuity of time can be conclusively resolved with the original Einstein theory of time as space-time, or local time.

There is obviously fear in some quarters that time cannot be something we invent by ourselves. Of course, if 'there is no longer a universal time' we have to find out how we get our time.[215] However, nobody is claiming that man invented the whole of time. Rather we have found that we invented how to quantify time by linking the natural sense of time as duration in the mind to external cycles. This sense of duration of

---

[215] It is not often realised that philosophy is of great importance to science; and, as an example, this is the sort of thing philosophers do behind the scenes to make their suppositions indispensable to science in general; for the philosophers service every branch of science. The phrase 'survival of the fittest' from biology which has passed into general usage in science and linguistics, was coined by a philosopher, not Darwin. All the sciences need philosophical interpretations. In the quotation above from Professor Dingle, he was saying this very strongly in respect of physics; but all the sciences need the same sort of assistance from philosophy, including mathematics and logic.

anything is obviously connected with the memory mechanism for the retention of images and concepts in the mind.

Let me stress again, and more strongly, that the sense of time is duration in the mind. In his Mathematical Theory of Relativity, Professor Eddington made this absolutely clear, as quoted above; and we have got to take that view seriously because the theory of time outlined in this book is based on relativity. Unfortunately the mental sense of duration is not enough. It cannot give time for general use because it is private. The word 'time' is meaningless until it is objectively quantified. We need time in units to apply to the external world---i.e. to mechanise in the clock for general use, so as to be able to tell 'How much time' at a glance---see Appendix I above. This is achieved with external cycles, the most basic of which is the earth-year out of which all other units of time are derived with mathematics. And it is maintained that this is in complete conformity with the Einstein notion of time, and therefore incontrovertible. Above all, it is one of the means by which we can logically solve the problems of the passage and continuity of time.

For now, we are told in all earnestness by some commentators that relativity is not properly understood. This may be so. But actually relativity is only a theoretical system, a suggestion. It is based on the suggestion that physical reality is not homogeneous but fragmented, and therefore subject to different natural laws. This applies to both special and general relativity. Bertrand Russell called it 'a logically deductive system'. In plain language, 'a new philosophy of physical reality' so logically structured that it demands attention, respect and serious study. And these Einstein has certainly achieved. With Einstein alone we are not talking about genius but a godlike intellectual phenomenon never seen on this planet before; he reconstructed the world of physical reality single-handed, that is the reason he is indispensable to both scientists and philosophers.

So Bertrand Russell was absolutely right. Einstein's system is a new logic of physical reality, and it works. But theoretical physics is most unlike the physics we apply in laboratories. Ordinary physics is much more like chemistry; it has consequences. The Nobel Committee was right to award Lord Rutherford the Prize for Chemistry, even though he regarded himself as a physicist, who had rather cheekily claimed that "all of science is either physics or stamp collecting"!

In theoretical physics there are no obvious consequences, so it is difficult to judge the merits of suggestions. Instead, when we get a new theory in advanced physics (rightly or wrongly), three things will happen. I mean, all three will definitely happen in succession, whatever may be the merits of the new proposal. First, we will get interpretations of the basic theory proposed in such complex settings (or confused formulas) from rival theorists that the debate just has to go on; nothing will be settled in the meantime. But because there are no consequences, nobody will get hurt, no machinery will fail to function; avoidable calamities will not occur. The rains will not stop; the sun will not dim.

The most recent example was the eather debacle (or debate). Secondly, we will get accusations and counter accusations of misrepresentations and misunderstandings. The third possibility (because philosophers share with theoretical physic one subject-matter, being the determination of physical reality), will be philosophical interpretations to arrogate the almighty right to shame and discredit some of the factions in the debate, only for philosophers of different schools to turn the tables---and so the debate will be carried on and on. These philosophical discourses are often pretty profound, giving several intelligent interpretations without being able to settle the argument one way or the other. Strangely, that is how we eventually acquire our knowledge of the external world, sometimes referred to as the practice of 'academic freedom'. That is what happened to Plato. And that is what is happening to Einstein as he has come to replace Plato, in fact, to make his basic suggestion redundant, if not completely false, due to the quantum theory.

A careful examination of what has happened to Einstein's theory of time so far betrays elements of all three conditions. First, we are told that 'most definitely' due to Einstein's analysis of 'Order and Simultaneity' there simply is no 'standard or absolute time frame in the universe'. ('Time Frame' or 'Time Reference' means the same thing. It means the logical criterion of validity.) This is generally accepted as true; for it is reinforced by the Lorentz time dilation and local time concepts.

However, it implies that time in the abstract is utterly indefinable, as I have shown above with discussions about the earth-year. The year is indefinable; other time units in use on earth are defined in reference to the year. But the year on its own is logically indefinable. Again, all time units, down even to the caesium units, are based on the earth-year; they are meaningful only as related to the year; but like the years, on their own (that is in the abstract), none of them can be logically defined. How long, for instance, is a second in logic without reference to something else? The result is that we all have to use the clock, or clocks, based on the earth-year. By this theory of time (as quantified time), the human intellect is built upon the concept of "points and instants". Instants do not exist independently in nature. Only points do; they had to be discovered by man, but they do exist in nature independently—for example, trees constitute points. Before we learned to put points on paper, we could see that trees dotted the landscape. Thus points constitute the basic instrument of human thought, especially in mathematics from which all the sciences spring. The instants arise from the act of 'consciously' and 'purposely' moving from point to point, confirming the Russellian notion that time is 'relation between points'. Hence quantified time is human in origin, except that the internal sense of time (as duration of anything in the mind) must be recognised as making a psychological contribution to the invention of quantified time in that the external cycles used for quantified time (the years, for instance), have to have psychological anchors (meanings) which are the sense of duration of anything in the mind.

Secondly, in the absence of a standard time frame, what does it mean to claim that time intervals in a moving frame are shorter---shorter as against what kind of standard or universal time? What time intervals are they compared with since there is no standard time frame? (Note that you cannot say they are shorter as compared to other clocks outside the moving frame; that will bring in the Einstein theory of frames, as I will discuss presently.)

So we all, in the end, have to resort to using the clock or clocks based on the earth-year. Yet if we use the clocks then it is not correct to claim that time intervals in a moving frame are shorter; they are not naturally or normally (in its proper setting) shorter or longer; they are normal to that frame, or to its natural frame. The moving clock may only seem 'different' as viewed from the outside; but if that is the case then there is no puzzle.[216] The time of the moving frame is not 'our' time; and it is

---

[216] Otherwise it is difficult to see how the behaviour of one clock can affect all

not queer to its natural environment or setting. It is a strange phenomenon to those looking in from the outside, in breach of the Einstein theory of frames. In fact, it is irrelevant to anybody but those in the moving vehicle only.

The whole idea of studying other frames from the outside is fraught with difficulties; it can never be an exact science since the standard postulates that make our system work (and make it what it is) might be inapplicable outside our frame, or planet.[217] Speculations into other frames from our frame have been responsible for all the bizarre suppositions about time and space-time from mathematicians and cosmologists in general relativity. I don't think that kind of enterprise is justifiable, especially when it leads to theories that space-time may be infinite in its timelike directions. Space-time cannot be infinite because it is necessarily discrete---the year, for instance, is not infinite. It is only one; all other units of time derived from the year are also discrete and individual. The proper way to think of time as space-time is that its units

---

time, human physiology and even the material contents of atoms, e.g. muons. If time is defined as the passage of existence in consciousness, how can the behaviour of one clock affect it for all of us? There is still a lot of religious beliefs about time. Time dilation is one of them, so sweet to the religious in science because they can claim that "it is a unique mystery about time predicted by Einstein". In fact, it is not a mystery, let alone predicted by Einstein: he rather solved the little problem with his theory of frames—i.e. the dilated clock belongs to another frame to which it is running normally.

[217] I think one implication of this is that the laws of physics, or some of them, would differ from ours at least in some parts of the cosmos, if not all over. Einstein was really a very strange genius in physical thought. He introduced the notion of postulates for natural laws in frames. This idea may go very far indeed in the cosmos at large.

are in perpetual procession (one year or second following another) to make time seem continuous; as such time can never be infinite.

Nothing illustrates the confusion about time in physics as a result of relativity and how it is misunderstood by scientists than the story of muons. By normal logic they should not last long enough to reach the earth; but they do. With the use of formulaic mathematics and concepts, physicists explain this by saying special relativity provides the answer as follows: the speed of muons is so great that their internal clocks slow down. Using the theories of time dilation and the so-called twin paradox based on it, it is assumed that as the muons gathered speed and their internal clocks slowed down they aged less and thus are able to last long enough to reach the earth. To a logician or philosopher who understands relativity, this is so laughable as to choke him. It is really the best example of the confusion in physics about time in relativity. (1) Time dilation has nothing to do with the muons and how they behave, since time does not dilate internally. Lorentz found that a moving clock would be seen by outsiders as running slowly; but internally those carrying the moving clock would notice absolutely no difference in its performance. Einstein explained this with his theory of frames---the moving clock is in a different frame. There is no logical mechanism for this kind of episode to be able to control time per se. All other clocks would not run slower or faster; and since there is no such thing as an absolute time frame, or standard time, by which all other clocks can be compared, the moving clock's performance has no relevance at all in physics, because its carriers would notice no anomaly; and those outside who notice any anomaly should mind their own business since it is not their time. (2) The idea that muons have internal clocks is based on the Minkowski theory of space-time, where space and time are assumed to constitute one entity; and therefore the reasoning goes that, since the muons occupy space, and all space is space-time, they have their own internal clocks to keep or measure time for them. Again, any logician will describe this as nonsense; for after all, the Minkowski space is known to be fictitious and arbitrary with absolutely no logical validity.

The basic idea in Time Dilation, which these writers rely on, is easily disproved thus: we know there are (roughly accurately) specific times by our normal clocks for the occurrences of certain events on this planet. Let us use Sunrise and Sunset for illustration. If Sunrise is usually 6 am, and Sunset is roughly 6 pm, as they are in some countries in the Tropics, it is inconceivable that a moving clock can force or influence these times to become 7 am, and 7 pm, on the planet all over just because one particular clock somewhere is running an hour late. "The dilation of time as a measure of moving clocks" can in no way influence all time per se on the planet.[218] It affects the performance of only one clock. Clocks are manufactured to reproduce specific time units, usually in seconds. If a particular clock, for whatever reason, is running erratically, there is no

---

[218] In any case, the quandary was solved with Einstein's theory of frames, as previously mentioned. Referring to it as if it were some kind of strange metaphysical phenomenon we do not understand, is part of the religious reaction to many aspects of relativity and time. Einstein did not define time in metaphysics. He merely pointed out that it must be added to the study of phenomena as a distinct co-ordinate in its own right. He also noted that by the analysis of simultaneity it just cannot be possible that time is absolute that generally permeates the cosmos and the same everywhere in the Newtonian sense. The interpretations of these ideas were left to philosophers, who have noted that Time Dilation was resolved by Einstein when it led him to the discovery of special relativity to the effect that the universe consists of fragments, each with its own natural laws. Our natural physical laws are influenced by the two postulates he gave us, from which we learn that Time Dilation occurs in a different frame.

logical mechanism for its behaviour to affect all other clocks on the planet.

The reader will have noticed that the name of Lord Bertrand Russell comes up regularly in all discussions of relativity's interpretation. It is inevitable. Russell was highly respected by Einstein, and for very good reasons. He was the world's greatest philosopher at the time. He was also a great mathematician and logician of genius. A most attractive writer, who won the Nobel Prize for Literature, he wrote about every subject in philosophy, including novels to illustrate moral points. When relativity was announced, he abandoned many of his most cherished ideas as wrong without shame or even mild embarrassment. He was candid and honest in the most adorable way, completely dedicated to the truth no matter how it reflected on his own beliefs. Russell probably had no certain beliefs other than the pursuit of the truth wherever it took him: via science, logic, mathematics or plain common sense, and linguistics. If he was certain that teaching mathematics to people from the cradle could save the world, he would have advocated that as his philosophy.

Concerning relativity specifically, in the later editions of his little book "Problems of Philosophy" he denounced his original philosophy as expressed in the book because of Einstein's theories, joking that whoever wrote the original ideas must have been a monkey, but nobody should suppose that the monkey looked, even remotely, like himself! No great philosopher has ever made such a confession; often associated with rulers, they all wrote imperious edicts as if they had discovered the final truth in logic and metaphysics.[219] Indeed, Russell later called his

---

[219] No surprise, then, that Russell later put them in their deserved places (mostly

Fellowship dissertation "somewhat foolish" for the same reason, namely, the geometry used by Einstein had made his discussions of the foundations of geometry completely wrong, and he was happy to admit it and adopt the new Einstein theory. He wrote one of the best interpretations of relativity, still in use, under the title "ABC of Relativity". His book "The Analysis of Matter" can be divided into two. One section is about relativity; the other is mainly about his joint theory with A. N Whitehead to the effect that the world of sense is a construction, not an inference. Yet even this can be traced to relativity, since Einstein made man the observer part of the observed, meaning that man contributes something to the nature of physical reality---i.e. to help with the construction of that reality---and the book was published long after both special and general relativity. It is a moot point.

---

of dishonour) in his monumental *History of Western Philosophy*. One complaint is that he never even once mentioned the name of Wittgenstein in this great book. The reason came from his contemporary, Sir Karl Popper---it was because, "In the long history of philosophy there are many more philosophical arguments of which I feel ashamed than philosophical arguments of which I am proud...Russell saw these things in that light, and so did I..." (From, *Modern British Philosophy*, By Bryan Magee, Secker & Warburg, London, 1971.) In 1959 Russell published his book, *My Philosophical Development*, in which he said he eventually had to reject Wittgenstein because he was talking 'logical mysticism' which was anathema to his basic nature. Of course he was right. Correctly defined, logical mysticism includes religion, mysticism and unscientific gibberish, all dressed-up to look like valid logical reasoning with a variety of linguistic trickery. Many aspects of philosophy in Oxford and Cambridge (and elsewhere) remain stuck in this kind of intellectual mud.

REFERENCES

With so many Footnotes, citing dozens of other books just to display learning is not my style. The books and papers cited below are the unavoidable ones for the rational discussion of time. I've used many of them in other books because my theory of time has not changed in fifty years!

ALBERT EINSTEIN (1879-1955) ---SPACE TIME, an article in the 1926/27 (13th) edition of the Encyclopaedia Britannica. Also, RELATIVITY, in the same edition.

---NATURE No. 106, 782, (1921), almost the whole issue was devoted to the confirmation of Einstein's new theory of gravity.

---The Meaning of Relativity, Princeton University Press, 1966.

---The Evolution of Physics, (With Leopold Infeld) Cambridge 1838.

---RELATIVITY, Routledge Classics, London and New York, 2001.

HERMANN MINKOWSKI (1864-1909) ---He first mentioned his supposition in a lecture in cologne, known as Raum und Zeit (Space and Time) Cologne 21st September, 1908.

--- Herman Minkowski AdP 47, 927 (1915)

---Herman Minkowski, Goett. Nachr., 1908 p53. Reprinted in Gesammelte Abhandlungen von Herman Minkowski. Vol. 2, p352. Teubner, Leipzig 1911.

The Origin of Secular Time

BERTRAND RUSSELL, FRS (1872-1970)---Our Knowledge of the External World, George Allen & Unwin, 1922.

--- Mysticism & Logic, George Allen & Unwin, 1976: a collection of important essays first published in 1917.

---ABC OF RELATIVITY, George Allen & Unwin, 1958 (recently revised by Professor Felix Pirani---first published in 1925.

---History of Western Philosophy, George Allen & Unwin, 1946.

---My Philosophical Development, George Allen & Unwin, 1958.

---The Analysis of Matter, George Allen & Unwin, 1927.

MORRIS KLINE: Mathematics in Western Culture, Allen & Unwin, London, 1954.

SIR ARTHUR STANLEY EDDINGTON, FRS (1862-1944)

---The Expanding universe, University of Michigan Press, Ann Arbor, 1933

---The Combination of Relativity Theory and Quantum Theory, Communication of the Dublin Institute of Advanced Studies, Dublin, 1943.

---The Mathematical Theory of Relativity, Cambridge, second ed. 1930.

---The Nature of the Physical World, Ann Arbor, Michigan, 1958.

---Philosophy of Physical Science, Cambridge, 1949.

Samuel K. K. Blankson

---The Theory of Relativity and its Influence on Scientific Thought, Oxford, 1922.

---Space, Time and Gravitation, Cambridge, 1920.

SIR JAMES JEANS, FRS: Physics and Philosophy, Cambridge, 1942.

---The Mysterious Universe, Cambridge, 1930.

---The New Background of Science, Cambridge, 1933.

PROFESSOR A.N. WHITEHEAD: The Concept of Nature, Ann Arbor, Michigan, 1957.

---Science and the Modern World, Cambridge, 1922.

---An Inquiry Concerning the Principle of Natural Knowledge, Cambridge, 1919.

---Nature and Life, Cambridge, 1934.

---Process and Reality: An Essay in Cosmology, Cambridge, 1929.

---Essays in Science and Philosophy, Rider & Co., London, 1948.

---The Principle of Relativity, Cambridge, 1922.

Professor BANESH HOFFMANN: The strange Story of the Quantum, Dover Pub. Inc. New York, 1959.

Professor STEVEN F. SAVITT (ed.) Times Arrows Today: Recent Physical and Philosophical Work on the Direction of Time, Cambridge, 1995.

CHARLES A. FRITZ: Bertrand Russell's Construction of the External World, Routledge & Kegan Paul, London, 1952.

The Origin of Secular Time

Professor JEREMY BERNSTEIN: Albert Einstein and the Frontiers of Physics, Oxford, 1996.

Professor RICHARD FEYNMAN: Lectures---The Character of Physical law. MIT Press, 1967. There are several volumes of the Feynman lectures and they are all worthy of serious study.

Abraham Pais, "Subtle is The Lord: The Life and Science of Albert Einstein", Oxford, 1982. Professor Pais has methodically provided details of almost all the original papers relevant to relativity. His list is so exhaustive I don't know of a better one anywhere.

WHAT REMAINS TO BE DISCOVERED, By Sir John Maddox, A Touchstone Book, Simon & Schuster, 1999.

INDEX

F

G

H

www.ingramcontent.com/pod-product-compliance
Lightning Source LLC
Chambersburg PA
CBHW071411180526
45170CB00001B/60